Time Matters

Dedication

To all the University of Sunderland geology students that I have been fortunate enough to teach. The late Professor Mervyn Jones, who encouraged me to open up my geological horizons, and the late Barbara Leddra, who sadly died before this book was completed – they are both greatly missed.

Time Matters: Geology's Legacy to Scientific Thought

Michael Leddra
Forefront Teaching Resources Ltd
Washington, Tyne and Wear, UK

WILEY-BLACKWELL
A John Wiley & Sons, Ltd., Publication

Blackwell Publishing was acquired by John Wiley & Sons in February 2007. Blackwell's publishing program has been merged with Wiley's global Scientific, Technical and Medical business to form Wiley-Blackwell.

Registered office: John Wiley & Sons Ltd, The Atrium, Southern Gate, Chichester, West Sussex, PO19 8SQ, UK

Editorial offices: 9600 Garsington Road, Oxford, OX4 2DQ, UK
The Atrium, Southern Gate, Chichester, West Sussex, PO19 8SQ, UK
111 River Street, Hoboken, NJ 07030-5774, USA

For details of our global editorial offices, for customer services and for information about how to apply for permission to reuse the copyright material in this book please see our website at www.wiley.com/wiley-blackwell.

Library of Congress Cataloguing-in-Publication Data

Leddra, Michael.
Time matters : geology's legacy to scientific thought / Michael Leddra.
p. cm.
Includes bibliographical references and index.
ISBN 978-1-4051-9908-7 (cloth) – ISBN 978-1-4051-9909-4 (pbk.)
1. Geological time. 2. Earth–Age. 3. Sequence stratigraphy. 4. Historical geology.
I. Title.
QE508.L43 2010
551.7′01–dc22

2009053117

ISBN: 978-1-4051-9908-7 (hbk) 978-1-4051-9909-4 (pbk)

A catalogue record for this book is available from the British Library.

Set in 10.5/13 Minion by Toppan Best-set Premedia Limited
Printed and bound in the United Kingdom

1 2010

Contents

Preface

This book is designed to give the reader an insight into the historical background behind some of the concepts we use in geology today. Due to the immense diversity of the subject matter now available to geologists and other Earth scientists, it would be impossible to cover all aspects of the subject here. I have therefore chosen to look at some of the most important or controversial topics that have helped shape modern geology.

The information contained in this book cannot be taken as complete, but is merely a starting point from which you can gain some understanding of geology and Earth Science and what lies behind some of the theories and concepts on which they are based.

The subject matter has been broken down into a series of specific topics, each forming a separate chapter. This process alone is difficult and somewhat arbitrary; as you will see that many of the topics interact and overlap without clear-cut boundaries. Equally, many of the characters who helped develop our concepts and practices crop up repeatedly. This forms one of the first important lessons within the science that we now call geology. In the past, scientists (a term which itself is quite new) were all-round natural philosophers: they studied the natural world as a single subject.

It is therefore important when reading this and other Earth Science books not to fall into the trap of treating each of the topics in isolation. It is also important to treat each of the ideas presented and the characters involved with due care: you must view them in the context and time in which they lived. For this reason, I have included birth and death dates for most people, to help you place them in history. So many geological and other science histories presented in books over recent years have been written with the beauty of hindsight, without remembering that later scientists had additional information to hand that the originators did not.

The book itself is a mixture of concepts and specific information. It can be used to provide a background to some very interesting ideas, both old and new, which have shaped and are continuing to shape our view of the world in which we live. To help you navigate through the ideas and personalities, the book contains three different types of text boxes. One type contains discussion points that I feel are worth thinking about. These also include either my comments or those of other people, which help set a scene, act as a prompt, or highlight specific points or views. I make no apology for the fact that a number of these come from other books, many of which are thought provoking. Many of these books have a very readable way of presenting concepts and ideas that some textbooks often fail to match. These boxes are designed to help you identify a train of thought that you might otherwise overlook, or to trigger a reaction. Other boxes contain background and additional information. I have also included many quotes from other books to provide a better understanding of the different ideas, characters, or the historical context of some of the subjects covered. A third type of box – titled further reading – has been used to highlight particularly interesting books, or parts of books that could help to develop your thinking or provide additional information.

If you would like to find out more about any of the ideas covered, I have included a list of some of the books that I used whilst writing this. It is not an exhaustive or even a comprehensive list, rather it contains many books that can be found on the popular science shelves in bookshops rather than just textbooks. After all, this is not meant to be just a textbook, but rather it has been written in a way that most people, from those with a geological background to others who just have an interest in Earth Science, will find useful. The book may also be of interest to people fascinated in the history and development of science. Obviously, however, some background knowledge of the subjects covered would help you to grasp what is being presented.

The original material in this book was designed to help teach university geology and geography students to think about the historical basis of their subject. It was also designed to encourage them to question their subject and to have a better understanding of the philosophy of science and how Earth scientists should make use of scientific methodology. One of the other primary roles of the original text was to teach students to look at evidence in their subject, and to put it into context with regard to current-day models. This goal remains. Science affects all of us in our everyday lives, even if we do not realize it.

From a geological point of view, it is essential that no one involved in our subject should take it for granted. Equally, I think it is fair to say that if we want people to appreciate why we think geology and Earth Science is so important, we should understand where our subject has come from and how it has developed. As you will see, geology and Earth Science have a very long history. The way in which they have developed has often been haphazard, being driven by different motives – some of which have not always been completely honourable. This is a good lesson to learn. Often our science, like many others, has been driven by personality and power rather than scientific reason and process. As you read this book, I hope you will find it interesting to see how geology and Earth Science have developed: they have sometimes lurched towards one view rather than another and then swung back again, or gone off in an entirely new direction. Whichever way this has happened, it is interesting to discover that overall with the checks and balances of the scientific method, these have balanced themselves out. This is an important lesson for the future. Sometimes ideas, which gain a "bandwagon", lead us up the garden path. However, it is satisfying to know that eventually views change for one reason or another and the subject progresses in a different direction.

Some readers may wonder why I have included ideas from Intelligent Design (ID) and "creation science" in a book that focuses on important ideas in geology. We can choose to ignore such ideas as fanciful, non-scientific, or attempts to fit physical evidence to a narrow set of ideas by showing that they are based on partial information or skewed examples. However, we risk, and it is a real risk, that people outside of the physical sciences and the public in general, may be swayed by such ideas. We should not be surprised when people take on face value the things they read, see, or are told by "experts", without understanding the validity or credibility of the information being presented or the people involved. We must be prepared to show, in a clear and understandable way, the shortcomings of such ideas, as it is becoming increasingly obvious that we cannot simply dismiss their ideas, even though many of them have been repeatedly disproved. It is astounding that many ideas rejected as long ago as the 19th century are still being regurgitated as new. Whilst many Earth scientists are engaged in publishing in learned journals, other people are running riot with our subject.

The Geological Society of London, the oldest national learned society for the Earth Sciences in the world, which "embodies the collective knowledge of nearly 10,000 Earth scientists worldwide", takes the issue seriously and published a statement during the United Nations International Year of Planet Earth, part of which has been included below:

> Approved by Council 10 April, and published 11 April 2008
>
> This Society upholds the right of freedom of belief for all. The freedom scientists enjoy to investigate the nature and history of the Earth is the same freedom that allows individuals to believe – or not – in a deity.
>
> Science's business is to investigate the constitution of the universe, and cannot pronounce on any concept that lies "beyond" Nature. This is the meaning of "agnostic", the word coined by former GSL President Thomas Henry Huxley, to describe a scientist's position of being "unable to know". This Society has therefore long operated according to the view that religion and science only become incompatible with each other when one attempts to trespass upon the domain of the other.

> The idea that the Earth was divinely created in the geologically recent past ("Young Earth Creationism"); attempts by Young Earth Creationists to gain acceptance for what they misrepresent in public as corroborative empirical evidence for this view ("creation science"); and the allied belief that features of the universe and of living things are better explained as the direct result of action by an intelligent cause than by natural processes ("Intelligent Design"), represent such a trespass upon the domain of science.

Finally, why is this book entitled "Time Matters"? It has often been said that the single most important contribution that geology has made to science is the establishment of the geological time scale and the age of the Earth. Nevertheless, I hope that by the time you have finished reading the book you will agree that geology has provided, and is still providing, far more than this. In particular – with the present focus on climate change – geology and the geological record has much to teach us about the Earth's climate and its effect on life: they are far more significant than the limited climate records that we hear so much about, that have only been collected over the past 150 years or so.

Michael Leddra
2010

Acknowledgements

This book is the result of a long and often convoluted process that began back in the 1980s. Whilst studying geology and geography at Kings College, University of London, I was fortunate to be taught by many fantastic lecturers, to one of which, the late, great Professor Jake Hancock, I owe the origins of this book. In one of his modules, he taught us about the history and politics of geology, about those who were part of the "in crowd", and those that were not, and their influence on the development of geology. The value of this has stayed with me and later, when I became a geology lecturer, it was incorporated into my teaching so that future geologists should know and understand where our subject had come from so that they would be able to see where it was going; I also wanted them to be able to tell other people about our history. Therefore, if any of them happen to read this, I am sure they will recognize at least the core of the book and I hope that it will bring back some interesting memories.

I would also like to thank three of my ex-colleagues, Andy Lane, Bill Scott, and Richard Lord in particular, who I was fortunate to work with at the University of Sunderland. These three geologists – much loved by all of our students – together with Iain Garfield, taught me more than they could imagine. Their enthusiasm and encouragement will always be with me. To be able to work with them in the field was a pure delight – we really did have fun, didn't we. A big thank you goes to Rev Peter Hood for his help and advice and to my very good friend, Rob Jones, for all his support and enthusiasm – especially for volunteering to read early drafts of the book, together with John Ball, Janine Marshall, Alison Machin and Karen McMullen. A special debt of gratitude must also go to Ann, my wife. who read numerous versions of this before I could pass it on to my niece Alexandra, who helped tidy it up.

I would also like to thank the enthusiastic members of the University of Sunderland Life-long Learning groups with whom I have been teaching and discussing this information over the past year – your questions, thoughts, and comments have been a real encouragement. Sincere thanks must also go to Ian Francis, Delia Sandford, and Kelvin Matthews at Wiley's for all their help and support during the preparation of this book.

Finally, I would like to say a big thank you to my family, who have had to put up with years of being battered by geology. Philippa, Stuart, and Adam, I am sure that you knew it was worth it, especially when you had to read some of this. As I always used to say to our students, "this is character building" and "remember, this is nice, we're having fun"!

Introduction

By way of an introduction, the following appeared in the preface of the County Durham Geological Conservation Strategy document:

> As far as is known, the word "geologia" was first used by Richard de Bury, Bishop of Durham from 1333 to 1345, who was a scholar and a tutor to King Edward III. He was a lifelong book lover and collector and wrote a book in Latin entitled "The Philobiblon" concerned with book collection and preservation amongst other subjects. In this book laws, arts and sciences are discussed and a new special term is introduced – geologia or Earthly science.

Looking and thinking about rocks is far older than this.

It appears that the following poem, written around 2,000 BC, refers to someone who may now be considered a geologist or miner:

> There is a mine for silver and a place where gold is refined.
> Iron is taken from the earth, and copper is smelted from ore.
> Man puts an end to the darkness; he searches the farthest recesses for ore in the blackest darkness.
> Far from where people dwell he cuts a shaft, in places forgotten by the foot of man; far from men he dangles and sways.
> The earth, from which food comes, is transformed below as by fire; sapphires come from its rocks, and dust contains nuggets of gold.
> No bird of prey knows that path; no falcon's eye has seen it.
> Proud beasts do not set foot on it and no lion prowls there.
> Man's hand assaults the flinty rock and lays bare the roots of the mountains.
> He tunnels through the rock; his eyes see all its treasures.
> He searches the sources of the rivers and brings hidden things to light.

This text comes from the Bible, more specifically the Book of Job, chapter 28, verses 1 to 11 (NIV version), which is thought to be one of the earliest books of the Bible to be written.Geology is therefore an area of the natural sciences that humans have investigated, theorized over, and exploited throughout his history.

Georges Cuvier defined geology, as we generally know it today, in the early part of the 19th century, at the request of Napoleon Bonaparte as part of a review of the sciences following the French Revolution. A similar definition was introduced at almost the same time in Britain, with the formation of the Geological Society of London.

As a modern science, geology covers a large variety of subject areas, many of which have stronger ties with other sciences than traditional geology. Because of this, I feel that it is therefore even more vital that people taking an interest in the subject have a clear understanding of where our subject has come from, and where it is now, so that in the future we can avoid some of the problems that are outlined in this book.

I have often described the work of a geologist as being similar to having a jigsaw puzzle where many of the pieces are missing and someone has lost the picture. Geologists therefore have to be able to collect, interpret, and assimilate a wide range of data (which is often incomplete) to be able to come up with an answer – the picture. This means that as the ideas and the data available develop, the picture can be re-interpreted. It also means that many aspects of geology are fluid and we cannot say that we have the full and complete answer. This can cause problems, arguments, and discussions that help the subject to progress. Added to this, because geology is all around us, we have not found everything yet, and we do not know all the answers – it is one of those subjects in which anyone can still find something new. Just think about it. If you are walking along and find a rock with a fossil in it and then crack it open, you become the first living thing to see that fossil since it lived, possibly hundreds of millions of years ago. That is some thought. In addition, as we have not found all the fossils yet, the one you find might be the first of its type. Now that is an even more amazing thought.

Geology is a four-dimensional subject in which time is probably the most important of the four dimensions. Not only do we have to deal with an immensely long time scale (4.6 billion years) but with processes, which range in occurrence, geologically, from a mere instance to countless millennia.

The following extract is taken from *T. rex and the Crater of Doom* by Walter Alvarez. This gives a very good overview of the development of geology over the last 350 years, in which he asks important questions such as:

> What kind of a past has it been? Is Earth history a chronicle of upheavals, catastrophes, and violence? Alternatively, has our planet seen only a stately procession of quiet, gradual changes?

Alvarez points out that for around 200 years, geological thought was dominated by gradualism – the idea that everything happens slowly over long periods – that was a philosophical reaction to catastrophic interpretations of previous generations. As we will see in later chapters in this book, new evidence and re-interpretations are enabling geologists to move away from this restricted viewpoint.

Geology and Earth Science have been, and continue to be, at the centre of some of science's great debates, many of which have been based around the age of the Earth and the speed at which Earth processes operate. Consequently, I have written this book in a way that reflects the progression of these themes. The first two chapters focus on views and concepts of the age of the Earth together with historic and modern methods of dating rocks. These chapters are designed to provide the foundations on which the rest of the book builds.

Chapter 3 looks at the way in which the geological time scale was constructed. One of the interesting aspects of this process was the apparently haphazard way in which it was pieced together. It is essential that the reader appreciates the timing of this process – the basic geological time scale and the major geological units (which are still used today) were founded on the observations made by natural philosophers **before** the publication of Darwin's *On The Origin of Species* and Lyell's predictions of a significantly old Earth constructed by slow, steady processes that operated over long periods of time.

Chapters 4 and 5 look at the shift from catastrophic and largely biblical-based concepts of Earth history to the ideas of slow, long-term changes and then back to the modern idea that Earth history is a mixture of the two. Having focused on rocks and processes in the first five chapters, attention switches to concepts of the development of life on Earth in Chapter 6. This includes a discussion of the way in which science does or does not work. This is followed in Chapter 7 by the debate – which still rumbles on – between creationists and evolutions. This is by no means a dead subject, as you will see, as there is an increasing push to have both taught in our schools. To some extent, this was one of the driving forces behind writing this book; to try to clearly lay out in a balanced way, I hope, where we are in our understanding of the Earth and its life forms, for those not directly involved in the arguments.

In Chapter 8, I have detailed the historical development of a continuously moving Earth surface from the concept of Continental Drift to Plate Tectonics. Even though both ideas are fairly recent, you will see that neither had a smooth development in the scientific age in which we live. The final chapter attempts to put everything into context and includes a number of anecdotes, which hopefully serve as a warning that we do not necessarily have all of the answers – let alone know all of the questions – and that, no matter how good we think we are, we still make some of the same mistakes that others made in the past.

I hope that by the time you get to the end of the book you will have gained an insight into geology and Earth Science that you did not have before, and will be ready to do "your bit" to either take them on into the rest of the 21st century, or at least follow them with eager interest.

1
Geological time

1.1 Introduction

Why is geological time so important? It underpins everything we study in geology and Earth Science today and provides a framework for many other sciences. The age of the Earth and length of geological time have probably occupied human thought ever since we became conscious of our surroundings. For centuries, people have attempted to quantify and measure it with varying degrees of success. With a significant degree of certainty, we can now say that the Earth is around 4.65 billion years old, a figure that is, to most people, unimaginably long. Many "creation science" articles and books that talk about creationist stratigraphy, repeatedly claim that we use and misuse – in their view – geological time. So, is our perception of geology, Earth Science, the rock record, and geological time wrong?

Perception of time has changed significantly throughout history, depending on a variety of factors. Some of these will become apparent in the following chapters.

When we look at much of the controversy that surrounds geology and other sciences, we find that the perception and determination of time is frequently at the heart of the problem. However, why should this be an issue? The amount of time available for something to occur usually increases the possibility of it happening or the frequency at which it can take place. If time scales are short, changes and variations become more important; but as time scales increase, it is possible for the unusual to become, if not the norm, at least unexceptional. This is why time has been an almost constant battleground for centuries and why, even now, it plays an important part in how different groups of people think about Earth Science.

For this reason, it is worth spending some time reviewing the historical perspective of time and how, in the past, people have perceived and tried to determine the age of the Earth. This will give us an insight into how different views have developed or changed over the centuries.

Time Matters: Geology's Legacy to Scientific Thought, 1st edition. By Michael Leddra.
Published 2010 by Blackwell Publishing Ltd.

Discussion point

Before you read the following sections, consider your views on the following questions:

Why should the establishment of geological time be so important?
What are your perceptions of geological time and the age of the Earth?
Why should it be necessary to have some idea of the length of the age of the Earth?

1.2 The historical perspective

The Greeks and Romans identified many of their gods with geological processes. As early as the 6th century BC, the Greek philosopher Miletus thought that geological processes were the result of natural and ordered events, rather than the result of supernatural intervention. Equally, another Greek, Democritus, thought that all matter was composed of atoms and therefore formed the basis of all geological phenomena.

During the 4th century BC, Aristotle identified fossil shells as being similar to living seashells and therefore decided that as fossils are found on the land, the relative positions of the land and the sea must have changed in the past. He also felt that for these changes to have occurred, long periods of time would be required. One of his students, Theophrastus, later went on to write the first book on mineralogy – entitled *Concerning Stones* – which formed the basis of this subject through to the Middle Ages.

During the Medieval Dark Ages, people viewed the length of time that the Earth had been in existence as very short and, since the Earth had been made for humans, its historical time frame was man-based. It therefore had a beginning that was "not long ago, and ultimately, an end not far in the future". In other words, they had no real estimates of time but thought that the Earth was fairly young.

The idea that the Earth was only 6,000 years old was based on a combination of the six days of creation and Jesus' words: "one day is with the Lord as a thousand years, and a thousand years as one day", which is recorded in 2 Peter, chapter 3, verse 8, which leads to the age of 6,000 years (and 4,000 years before the birth of Christ). An age of 3952 BC was proposed by The Venerable Bede (672–735) and in 1583, Joseph Justus Scaliger (1540–1609), a French scholar, published *De Emendatione Temporum* in which he calculated the formation of the Earth to be 4713 BC (some publications include other dates, such as 3929 BC). He arrived at this date based on the combination of three known cycles:

1. The Solar Cycle, which refers to the 28-year cyclic behaviour of sunspots;
2. The Metonic Cycle, which refers to the 19-year period it takes for the same phase of the moon to occur in the same calendar month;
3. The Roman Induction, a cycle introduced by the Emperor Constantine for tax purposes, which has a period of 15 years.

Scaliger thought that the creation of the Earth must have been the first time that all three cycles coincided and he calculated that this occurred in 4713 BC, and would not occur again until 3267 AD. His date was used as the start date for the Julian calendar, which was introduced by Gaius Julius Caesar.

One of the first people to attempt to establish the age of the Earth, based on chronologies listed in the Bible, was Sir John Lightfoot (1602–1675). He was a vicar born in Stoke-on-Trent, who eventually became the Vice-chancellor of the University of Cambridge. Using the Bible as a reference, between 1642 and 1644 he decided that the Earth was formed at 9 am on the 26 October 3926 BC.

Shortly after this, James Ussher (1581–1656), who was Vice-Chancellor of the Trinity College Dublin in 1614 and 1617 and Bishop of Armagh in 1625, also calculated the age of the Earth using the Bible (Fig. 1.1).

Fig. 1.1 **James Ussher**

Ussher has usually been portrayed as a symbol of authoritarianism and religious dogma, with terms such as "rule of authority", "early speculation", and "foolish" frequently being used with regard to him and his work. It is also said that he "pronounced" or "announced with great certainty" his date for the formation of the Earth. However, Ussher was renowned as an eminent scholar in his time. In 1640, he came to England to undertake research, and between 1650 and 1654, he made a detailed study of the Old Testament in which all the generations of people that had lived since the creation of the Earth are recorded. Many books imply that he did this simply by summing their combined ages and then calculating that the Earth was formed on Tuesday, 23 October 4004 BC, at 12 noon (some texts give the date as either the 22nd, 24th, or 26th and also state that Ussher put the start at 9 am, but this is incorrect). Although many later natural philosophers and geologists poured scorn on this estimate, it should be noted that most geologists for the next 100 years did not envisage a time span that was significantly different. This might have been because they accepted his ideas or because they could find no better way to determine it. Ussher and his work are often viewed as science bound up by religion, but his was one of the first serious attempts to organize the ever-expanding amount of information about the Earth into a coherent story and time frame.

In his day, Ussher had a reputation for being moderate, willing to compromise, and a keen scholar. In 1650, he published the *Annals of the Old Testament*, in which he presented his data. This book, which contained his deductions for the timing of the origin of the Earth, was 2,000 pages long and could hardly be regarded as a minor or rushed piece of work.

He represented a major style of scholarship of his time, in which he – working in the tradition of research – took the best sources of information and evidence available to try to determine the answer to a specific question.

Discussion point

Contrary to the usual versions of Ussher's work, he did not "simply add up the ages and dates given directly in the Old Testament", but made a valid attempt to estimate the age of the Earth, using what he considered to be the best, most accurate, and faultless data he could find.

So, how did he come up with 12 noon on the 23 October 4004 BC as an exact time for the formation of the Earth?

1. The year of 4004 corresponds to six days of creation, where 1 day equals 1,000 years, which was a common comparison at the time.
2. Why 4004 and not 4000? It had already been established that there was an error in the BC to AD transition, as Herod died in 4 BC. The date could actually have been anywhere between 4037 and 3967 BC.
3. Why 23 October? This is based on the Jewish year that started in the autumn. He thought that the first day would follow the Autumnal equinox.
4. Why more than a month after the equinox? Dates at the time were based on the Julian system similar to that used today, except for one thing – they did not leave out leap years at century boundaries (i.e. divisible by 400). Thus, by 1582, the calendar had ten extra days (Pope Gregory XIII established a new calendar that has come to be known as the Gregorian calendar, in which Thursday, 4 October was followed by Friday, 15 October 1582). However, this was not adopted in Britain until 1752. Therefore, in 4,000 years there would be an extra 30 days.
5. Why midday? Ussher began his chronology at midday as "you cannot have days without the alternation of light and darkness" and as the Bible says, "in the middle of the first day, light was created".

Discussion point

In his book *Eight Little Piggies,* Stephen Jay Gould illustrates a very poignant lesson:

> How many current efforts, now commanding millions of research dollars (or pounds) and the full attention of many of our best scientists, will later be exposed as full failures based on false promises? People should be judged by their own criteria, not by later standards that they couldn't possibly know or assess.

This is an important point that should be taken into account when considering everything that follows.

In his book, *Revolutions in the Earth: James Hutton and the True Age of the World*, Stephen Baxter points out that:

> Even as Ussher was publishing his great work, doubts were raised. During the previous centuries Europeans had begun to travel the world. And they encountered cultures which had their own historical narratives, many of them contradicting the biblical account. The Chinese for example, mocked

the story of Noah's Flood, which was supposed to have occurred around 2300 BC. Chinese written history stretched back centuries before this date, and made no mention of a disastrous global deluge.

After his voyages of discovery, Sir Walter Raleigh (1552–1618) felt that the history of the Earth was considerably older than that envisaged by the Church. Because of this, he and his friends were frequently accused of atheism and heresy, even though when he wrote *The History of the World*, none of these views were expressed.

1.2.1 The march of the scientists

In the early 17th century, natural science was a very popular subject, with many people beginning to study Nature, rocks, and fossils. Many viewed the creation of rocks as the result of Noah's Flood, which followed the Bible's version of history and implied that the Earth could not be very old.

Even so, people like the famous physicist Robert Hooke (1635–1703), who was one of the founding figures in the understanding of earthquakes, did not believe that sedimentary rocks were the results of Noah's flood. Neither did he envisage a greatly extended time scale beyond that proposed by Ussher. He, like many other scientists at the time, viewed the geological history of the Earth as being increasingly violent the further you went back in time. Shortly after the Royal Society was founded by Charles II in 1660, Hooke was appointed Curator of Experiments and most of his geological work was published in 1705 under the title *Lectures and discourses of earthquakes and subterraneous eruptions, explicating the causes of the rugged and uneven face of the Earth; and what reasons may be given for the frequent finding of shells and other sea and land petrified substances, scattered over the whole terrestrial superficies.*

In his book *The Making of Geology: Earth Science in Britain 1660–1815*, Roy Porter reveals that following the formation of the Royal Society, there began a significant change in which wild speculation and theory was gradually replaced by detailed observation, comparison, and description. Fieldwork was also becoming an increasingly important component of natural history.

It was not until the mid- to late-17th century that people began to realize that rocks had not been laid down by Noah's Flood, but had a recognizable continuity, sequence, and distribution that meant that they also had a history. As Porter puts out, "there was a growing realization that strata were the key to Earth history".

Nicholas Steno (1638–1686) (Chapter 2), like Hooke, played an important role in establishing some of the fundamental principles of geology. He used his knowledge of fossils and stratigraphy to establish the geological history of Tuscany. By looking at the rocks around Florence, he determined that the area had been flooded twice. During the first episode, the older strata were deposited. They were then tilted and a second flood deposited more sediments on top of them. He rationalized what he saw with scriptural accounts and suggested that the first flood was that of the second day of creation and that the second flood was that of Noah. With this interpretation, he did not expect to find evidence within the rocks of an extended geological history.

Isaac Newton (1643–1727), the celebrated physicist and mathematician, calculated that the Earth was 6,000 years old based on the Bible, and what he considered other reliable sources of information (Fig. 1.2). As Jack Repcheck says in his book, *The Man who Found Time: James Hutton and the Discovery of the Earth's Antiquity*:

> All Christian churches, their clergies, and their followers – believed that the earth was not even 6,000 years old. This belief was a tenet based on rigorous analysis of the Bible and other holy scriptures. It was not just the devout who embraced this belief; most men of science agreed that the earth was young. In fact, the most famous of them all, Isaac Newton, had completed a formal calculation of the age of the earth before he died in 1727, and his influential chronology confirmed that the biblical scholars were right.

Fig. 1.2 **Sir Isaac Newton**

Fig. 1.3 **Sir Edmund Halley**

In 1715, English astronomer Edmund Halley (1656–1742) – who was also a mathematician, meteorologist, and physicist – suggested that the age of the Earth could be calculated from a study of the salinity of the oceans (Fig. 1.3). He thought that if he determined the salinity of the sea, and re-measured it after a period of time, he could calculate how salinity changed with time. From the expected increase in salinity, he felt that he could then back-calculate the amount of time it would take for fresh water to achieve the present salinity of the sea. It is not known whether he actually carried out such experiments.

He then published his theory, titled *A short account of the cause of the saltiness of the oceans, and of several lakes that emit no rivers; with a proposal by help thereof, to discover the age of the World*, in the Philosophical Transactions of the Royal Society of London.

Within this theory, he assumed that the oceans were originally composed of fresh water and that their salinity had progressively increased through the transport of dissolved materials by rivers that flowed into them. He used lakes as a further source of data and categorized them as either those that had inlets and outlets to the sea or those that only had inlets. He used these as examples of the way in which the salinity of the oceans could have formed, i.e. saline water flowing into them that could not flow out again, which would account for the gradual increase in salinity. Halley recognized that there could be several flaws in this argument. For instance, there is not a constant influx of materials, without loss through burial, which would lead to salt being trapped in sediments and rocks that would then prevent it from being dissolved in the sea. He did not, as is usually implied, set out to show that pervious estimates, such as those mentioned above, were wrong, but used similar arguments to those of Ussher.

He is often portrayed as a "hero of geology", because his "scientific methodology" provided a minimum (and not a maximum as has sometime been portrayed) estimate for the age of the Earth. This meant that the Earth was older than previously thought. He also recognized the existence of a second flaw: the oceans may not have consisted of wholly fresh water to start with. He argued therefore that his methodology would provide a minimum age of the Earth, because if there was any salt present in the oceans from the start, his method would reduce the calculated age. This is a crucial argument because he was concerned about Aristotle's ideas that the Earth and therefore time were eternal. This concept also made the idea of extreme events impossible, as his theory relied on a constant change in salinity. (See Chapter 5 for Halley's ideas on magnetism and the structure of the Earth.)

While he was the French Consul in Egypt, at the age of 35, Benoit De Maillet (1656–1738) wrote a book based on his observations and studies of Egyptian records that detailed the flooding of the Nile (Fig. 1.4). He noted that Carthage – a fortress on the seashore that had openings in its basement to let sea water in – stood 1.5–1.8 m above sea level. He estimated, from data recorded over 75 years, that sea level had dropped by 7–9 cm per century. He also noted that similar occurrences were recorded at Acre and Alexandra. De Maillet argued that this data provided evidence for a gradual decline in sea level of around 3 m a century. This could be projected back to a time when the Earth was covered in water and the ancient mountains "emerged from the sea" (a reference to a universal flood). Given the heights of the known mountains, it led him to estimate that the "Earth was immensely ancient", around 2,000 million years old. In order to evade censorship by the Church, he presented the theory titled *Telliamed: or Conversations between an Indian philosopher and a French missionary on the diminution of the sea.* Telliamed, which was De Maillet spelt backwards, was supposed to be an oriental philosopher. It was subsequently circulated clandestinely in 1720, but was not published until 10 years after his death.

Fig. 1.4 **Benoit De Maillet**

Discussion point

We find evidence of sea level change across the country, from flooded river valleys in Devon and Cornwall to raised beaches around Scotland and elsewhere (Fig. 1.5). Without the knowledge of sea level change that we now have, how else would we be able to account for these changes?

The above are examples of estimates of the age of the Earth based on the observation of natural processes. Others also produced estimates based on experiments and scientific theory. These include those by George Leclerc (Fig. 1.6).

George Louis Leclerc, Comte de Buffon (1707–1788), was the keeper of the Royal Zoological Gardens in Paris. As a Newtonian scientist – that is, someone who believed that everything in the universe was measurable and founded on mathematically determined laws – he followed De Maillet's model. He spent 10 years collecting and cataloguing the collections of the

Fig. 1.5 **Two examples of raised beaches, on the Isle of Portland, Dorset (top left and right), indicating a significant change in sea level, and the Isle of Arran (bottom left) with abandoned stacks and cliffs**

Fig. 1.6 **George Louis Leclerc, Comte de Buffon**

Natural History Museum and set about producing what he estimated would be a 50 volume encyclopedia. The first three volumes were published in 1749. This ultimately resulted in his 36 volume *Historie Naturelle* (Natural History), which was published between 1749 and 1789. This was the first naturalist account of the history of the Earth and the first account of geological history that was not based on the Bible. He only used observable or quantifiable causes to explain natural phenomena.

Another important publication was his *Époques de la Nature,* published in 1778. In this, he proposed that a comet collided with the sun, causing 1/650 of its mass to break away and form the planets.

He proposed that the history of the Earth could be divided into seven epochs, based on different stages of cooling, the first of which saw the Earth being composed of molten material. The first epoch included the formation of the Earth as an incandescent mass that, according to Douglas Palmer in his book *Earth Time: Exploring the Deep Past from Victorian England to the Grand Canyon*, he called "a 'vitriscible' rocky state as it cooled". It gradually cooled down during the second epoch and cracked to form high mountains. The third epoch saw the condensation of water and the initiation of rain. Organic matter then formed "spontaneously by the action of heat on fluid substances". The fourth epoch saw the emergence of a land surface from beneath the water. Palmer outlines Buffon's view of the historical development of the Earth that included a very interesting insight, given the time in which it was written:

> there was a general flood, which on retreat left fossil shells embedded in its sedimentary deposits. The large quadrapedal animals followed next and to Buffon their global distribution showed that the continents must have been joined as a single mass. The sixth epoch saw the continents separate and finally in episode seven mankind appeared.

Buffon's idea that there must have been a single large landmass that then separated had significant implications that he must have considered at the time (Chapter 8).

Buffon modelled the Earth using ten balls of mixed iron and non-metallic minerals in varying sizes that increased incrementally by 1/2 inch (12.5 mm) up to 5 inches (130 mm), which were made for him in ironworks that existed on his estate in Montbard near Dijon. These were then heated almost to their melting point and allowed to cool. From these experiments, he calculated that, given the size of the Earth, it would take 74,000 years to cool down from its molten state, in addition to 2,936 years for its initial consolidation. The Granite Mountains, he said, were the only parts of the original crust still visible. As the Earth cooled, after 50,000 of the 74,000 years, a rain of nearly boiling water began to fall that covered or nearly covered the entire surface. He also claimed that as volcanoes were only found near the seashore, they were the result of chemical activity in modern times, powered by steam from water seeping into the Earth.

Stephen Baxter notes that when Buffon calculated the age of the Earth as 74,000 years old, his work "created a great furore … The theologians at the Sorbonne condemned him, and Buffon dutifully retracted". However, he was not sincere: "It is better to be humble than hanged". Buffon continued his experiments, and rather than a time span of 74,000 years, he published estimates that were in the order of 3,000 years rather than the 3 million years that his unpublished material indicated he considered to be nearer the true age. This was, as Baxter explains, not "so much from fear of the religious authorities – by now he was too old to care – but because he thought the public wasn't ready for them".

Discussion point

Why do you think that there was such a difference between his published and unpublished figures?

Why do you think Buffon divided his history of the Earth into seven epochs?

In 1756, Immanuel Kant (1724–1804), the famous German philosopher, proposed that the Sun's energy was generated by the combustion of conventional fuel. In so doing, he estimated that, given the size of the Sun, it would burn up within 1,000 years.

James Hutton (1726–1797), studied law at Edinburgh, but preferred chemistry; and ended up studying medicine, which he finished in Leiden in 1749. He never practised medicine and after his father died, he ran his farm in Berwick. In 1768, Hutton moved to Edinburgh and joined a circle of eminent scientists and philosophers that included the inventor James Watt. In 1785, he published a paper on geology in which he summarized his new theory of the Earth. This was attacked in 1794 by Richard Kirwan, who was a follower of Werner (Chapter 3), and in response Hutton rewrote and expanded it into a full account titled *Theory of the Earth*, which was published in 1795 (Chapter 4 and 5), in which he proposed that the Earth was significantly older than previously thought.

Following the work of Buffon, Georges Cuvier (1769–1832), the Chair of Comparative Anatomy in Paris, looked at the fossils in the Tertiary rocks of the Paris Basin (Chapters 5, 6 and 7). Again, as with Hutton, there was no indication of a specific age but it was his view that the Earth had passed through "thousands of ages".

The 19th century is considered the "heroic period" of geology. By 1840, the stratigraphic column – the sequential order of rocks – as we know it today, was virtually complete, which allowed every rock on the Earth's surface to be allotted its relative time position in the sequence (Chapter 3). This led to a change of emphasis with regard to the problems of trying to determine geological time and the age of the Earth: with the establishment of the rock sequence people began to realize that they told a story – the historical development of the Earth.

In 1830, Charles Lyell (1797–1875) – who became one of the most famous geologists – published the first volume of his *Principles of Geology*. In 1867, he tried to calculate the length of geological time since the Ordovician Period (Chapter 3), based on an estimate that it would take approximately 20 million years for a complete change of molluscan species to occur. He noted that the fossil record indicated that 12 such changes were recorded since the start of the Ordovician Period, hence this would represent a time period of 240 million years. (We now know that this is actually closer to 500 million.) Lyell's estimate was totally unverifiable because there was no way of calculating absolute dates at the time, but it was one of the first estimates that talked in terms of hundreds of millions of years rather than thousands or even millions of years.

In 1850, Physicist Hermann Von Helmholtz (1821–1894) thought that the Earth had been formed by the collapse of material into its centre, which in turn converted gravitational energy to light and heat. Using this process as a basis, he estimated that the Earth was between 20 and 40 million years old.

Charles Robert Darwin (1809–1882) studied theology at Cambridge and became a keen geologist. When he published *On The Origin of Species* in 1859, one of the crucial factors that he needed for his theory of evolution to work was an extended Earth history. Lyell's estimation for the age of the Earth and his ideas of uniformitarianism provided the time scale he needed. Darwin estimated that he required a time period of at least 300 million years since the last part of the Mesozoic Era (Chapter 3) for evolution to work. This value is now known to be far too large, as most geologists consider this time period to be about 65 million years.

In 1878, the Irish geologist Samuel Haughton (1821–1897) introduced the idea that you could estimate the age of the geological record by adding together the thickness of all known strata. For over 50 years, this proved to be a popular way of estimating the age of the Earth.

In 1899, John Joly (1857–1933), like Halley, looked at changes in salinity to determine the age of the Earth. He thought that if he looked at the chemistry of rivers he could calculate how much sodium was being added to the oceans each year. He decided that if he could calculate the approximate volume of the oceans, he could estimate the amount of time necessary to achieve their present salinity. Allowing for salt that was blown back onto the land, he estimated that 90 million years had elapsed since fresh water had first condensed on the Earth's surface.

Discussion point

The problem with estimates based on changes in salinity is that the natural system in which it exists is very complex, and the estimates all assume that the oceans originally consisted of fresh water – the product of a universal flood formed by rain water. Another problem was calculating the volume of water in all of the oceans, something that is challenging today and would have been even more so in the 19th century. It is now believed that the salinity of the oceans does not vary with time.

William Thomson (1824–1907), better known as Lord Kelvin, became Professor of Natural History at the University of Glasgow at the age of 22 (Fig. 1.7). He was the co-discoverer of the Law of Conservation of Energy. Kelvin thought that if the Earth worked as a heat machine, it would be possible to determine how old it was. In 1862, he set himself the task of calculating how long it would take the Earth to cool down from its original molten state, and he continued to work on this idea, on and off, for 42 years. He also considered the age of the Sun and its rate of cooling. Based on the heat output from the Sun, Kelvin calculated that it had "shone down on the Earth" for about 100 million years. However, as its heat built up due to collision of smaller masses, it would have only generated a temperature hot enough to sustain life on the Earth for the last 20–25 million years. He also looked at the Sun's energy, which was thought to be generated by gravitational contraction. Due to its size, Kelvin thought that this process would make the sun fairly old – although he believed that it had only illuminated the Earth for a few tens of millions of years. He also thought that if the Sun's energy was more than 10% either side of its present value, life would then cease to exist on Earth.

Initially Kelvin set a wide margin of error, of between 20 and 400 million years, for his estimates of the age of the Earth. By 1897, he had refined these

Fig. 1.7 **William Thomson, Lord Kelvin**

down to 20 and 40 million years. He was a very influential scientist at the time and because his calculations were based on precise physical measurements, this rendered his estimate irrefutable. He looked at the uniform increase in temperature with depth in deep mines (as had Joseph Fourier nearly 100 years earlier) to calculate the Earth's thermal gradient; this indicated that heat flowed from the centre to the surface. He reasoned that, as the Earth was gradually cooling down, he could back-calculate the rate of heat loss to determine when it had originally been molten. Even without knowing precise details on the melting points of many of the Earth's rocks and minerals, his estimates were always less than 100 million years. Because of his reputation and scientific credibility, most geologists accepted his estimates – even if they had reservations about the relatively short time scale involved.

Kelvin continued to re-evaluate his ideas and recalculate the age of the Earth until he reduced his estimates down to only 24 million years, at which point various geologists – including Archibald Geikie, who had been studying ancient volcanism in the British Isles – refused to believe him. It is interesting to note that, using his methodology and the data we now have, we would still come up with 25 to 30 million years based on the assumption that the Earth is gradually cooling down.

Discussion point

One of the problems with such a short time scale is that it had serious repercussions for Darwin's theories of evolution. Even Darwin, in later editions of *On The Origin of Species*, was hesitant about talking of a long geological history, because of this evidence and presumably because of the influence of Kelvin as a scientist.

It is important to remember that the estimates above were made **before** the discovery of radioactivity and the realization that this was the source of the Earth's heat and it was therefore not cooling down.

Charles Doolittle Walcott (1850–1927), who is reputed to have had only one week of college training with Louis Agassiz in 1873, later went on to become a famous geologist with the USGS (United States Geological Survey). In 1893, he calculated, by using the total thickness of the known rock sequence in the geological record and typical rates of sedimentation, that the Earth was approximately 75 million years old.

Discussion point

Calculating typical rates of deposition has always been a difficult process, particularly during this period when people continued to find new sections around the world. Average rates of sedimentation vary not only for different periods but also within the same periods. When using rates of deposition, scientists must also take into account the rate of compaction of the sediments.

Even though they were using a completely different method for calculating the age of the Earth, the people who used deposition rates appear to have consciously (or subconsciously) stayed within the estimates of Kelvin.

In 1900, William Johnson Sollas (1849–1936), a geologist from Oxford, estimated geological time as being 18.3 million years.

1.2.2 The atomic age

After Henri Becquerel in 1896 (Fig. 1.8), Marie Curie in 1903 and others had discovered radioactivity, physicist Robert John Strutt estimated the amount of heat that was continuously being generated by radioactive

Fig. 1.8 **Henri Becquerel**

minerals in the Earth's crust (Fig. 1.9). From this, he showed that this accounted for the Earth's geothermal gradient and the apparent heat flow to the surface, without the need for the Earth to be cooling down. This meant that there were no longer any problems with having to stay within or question Kelvin's dates (Chapter 8).

In 1902, Ernest Rutherford (1871–1937) (Fig. 1.10) – a New Zealand born physicist – and Frederick Soddy – an English chemist – published the results of their first experiments with radioactivity. According to most sources, late in 1904 (some quote 1905), Rutherford suggested that alpha particles were released by radioactive decay and that this decay could be used to determine the age of rocks, a technique he named radiometric dating. His first book *Radioactivity* was published in 1904. The same year Rutherford presented the first radiometric date for a rock, a sample of pitchblende, based on the uranium/helium method, at the International Congress of Arts and Science in St Louis. Most sources quote the age as 500 million years, although one source also gives this as 700 million years. (Some sources imply that the original experiment was conducted by Sir William Ramsey.)

Fig. 1.9 **Marie Curie**

Fig. 1.10 **Ernest Rutherford**

In 1905, Bertram Boltwood (1870–1927) – a pioneering American physicist and chemist – developed a technique for determining the radioactive age of rocks that contained uranium. Using this methodology, he determined the age of 26 samples of rock, which produced dates between 92 and 570 million years. Unfortunately, his measurements were flawed and later tests would show that the rocks were in fact between 250 and 1.3 billion years old.

In 1907, Boltwood made the first attempts at constructing a geological time scale using the radiometric dates, the same year that Rutherford, whilst working with Hans Geiger at the University of Manchester, was developing a method of detecting and measuring radioactive particles using electricity. These and other discoveries led to the establishment of radioactive dating, as detailed in a paper written by Arthur Holmes in 1911. In 1910, he had dated a piece of rock from Norway at 370 million years which, as the rock had been formed during the Devonian Period, meant that for the first time he was able to give this geological time period an absolute date. He also calculated the age of some rocks from Greenland, which gave an average age of 3,015 million years.

Whilst Holmes was working at the Strutt laboratory, under the guidance of Robert John Strutt (1875–1947), he estimated that the Earth was 2.4 billion years old and produced the first absolute time scale, which he published in 1913 in *The Age of the Earth*. Douglas Palmer advises in his book, *Fossil Revolution: The Finds that Changed our View of the Past*, that this gave the "first really reliable estimate of a minimum age for the Earth". He then adds that, "Holmes went on to estimate that the origin of the uranium, from which the lead was derived, must be around 4,460 million years ago".

Holmes worked on radiometric dating over a period of 50 years, until he and others discovered which radioactive elements and their decay products (usually referred to as daughter products) could be used as reliable "atomic clocks" (Chapter 2).

In 1929, Rutherford discovered that two such radioactive clocks, uranium-238/lead-206 and uranium-235/lead-207, ran at different speeds. In fact, the uranium-238/lead-206 clock ran six times faster than the uranium-235/lead-207 one. This discovery meant that dates could be determined by comparing the growth of the two different types of lead. When it was finally confirmed that another lead, lead-204, was not derived from radiometric decay, this "primary lead" provided a third clock that could be used in the dating method.

Henry Norris Russell (1877–1957), an American astronomer, obtained a date of 4 billion years for the age of the Earth in 1921. Following the work above, Harrison Brown (1917–1986) and Claire Patterson (1922–1995), two American geochemists, measured the age of the meteorite that formed

Meteor Crater in Arizona and found it to be 4,510 million years old. Patterson also found that a number of other meteorites produced similar dates, which resulted in an average age of 4.55 billion years. This was close to Holmes' estimates for the age of the Earth and is close to the generally accepted age of 4.6 billion years used today. In fact, the oldest known minerals – zircons from Australia – have been dated as 4.4 billion years old.

Discussion point

Cherry Lewis, in her book *The Dating Game: One Man's Search for the Age of the Earth*, recounted the life of Arthur Holmes. In this, she sums up the problems that he struggled with and makes the following comments that could be used to summarize the development of many scientific ideas. She writes:

> Progress on dating the age of the Earth was slow, years, even decades went by without any significant advance being made. But science is like that. What is often not realized when a breakthrough finally occurs is that for years previously a few individuals had been diligently working in the background, thinking and writing about the problems, quietly and persistently pursuing their goal.
>
> ... having scrutinized in detail every analysis of uranium, lead, thorium, radium and helium that had ever been published anywhere in the world, now amounting to many hundreds, Holmes determined the age of the Earth.

She describes his work as a labour of love during which he spent years working without flashes of inspiration or miraculous discoveries, but just plain slow-going, hard work.

Stephen Jay Gould suggests that hindsight has frequently been used to pick out themes that seem to anticipate later developments. Similarly, passages are taken out of context in an effort to show how one writer or another had almost (but not quite) put together the conceptual framework that we accept today. Because of this, heroes have been made of those who had "forward-looking" theories and their "unmodern" ideas have been ignored. Rival theories were dismissed as obstacles to the development of science as they were only supported by conservatives who wished to retain traditional values and were deemed to be holding back the development of science.

It is interesting to note that frequently throughout *Time Matters*, we will see that it is often the ideas, or parts of the ideas of the "bad guys", which eventually have an important input to geology as we know it today. Conversely, those of the "good guys" frequently led to comparative dead-ends. It is also important to see that most of the time the development of knowledge is time-consuming, hard work that is often spread over a long period of time and involves blind alleys, false hopes, and misconceptions.

Lewis also includes Rutherford's description of the time he had "drawn the short straw" and had to present his new dates in front of Lord Kelvin during a lecture at the Royal Institution in London. She also describes another presentation at the British Association in 1921, in which Strutt was presenting Holmes work, where he "again tried to lay the spectre of Kelvin who still rose to haunt the assembly, and again put forward the arguments in favour of radiometric dating". She reports that William Sollas, Professor of Geology at Oxford, was apparently "overwhelmed" by the amount of geological time that was now available compared to that offered by Kelvin, stating that:

> the geologist who had before been bankrupt in time now found himself suddenly transformed into a capitalist with more millions in the bank than he knew how to dispose of, but perhaps understandably, he still urged caution and heeded geologists to substantiate the time being offered by physicists before committing themselves to the reconstruction of their science.

How and why have views of geological time and the age of the Earth been restricted by limited knowledge and/or scientific dogma?

1.3 Geological time and the age of Mother Earth

There are a number of different ways in which the enormity of geological time has been portrayed in a user-friendly way. These include a 24-hour clock, spiral, and linear diagrams. Each has their own merits but I think that a book title *Restless Earth* by Nigel Calder, that was published in 1972 to support a BBC television series, presented one of the best representations. This portrayed "Mother Earth" as a 46-year-old woman, which provided a time scale that was neither too compact nor too large to relate to. Calder's book was published whilst I was doing A level geology at school. I wrote this version of the time scale out and stuck it to the inside cover of my file, together with a postcard geology map of the UK from the Geology Museum in London. I still had it ten years later when I started my geology degree, as a reminder of why I first fell in love with the subject. The following is an adaptation of the *Restless Earth* time scale that I hope puts geological history in to context:

We could view Mother Earth as a middle-aged lady of 46 years old, where each of her "years" are mega-centuries (i.e. 1 year represents 100 million years).

Like so many people, details of the first seven years of her life are almost completely lost, apart from a few vague memories, and a few snapshots. Towards the end of her first decade, we have a better record of some of her deeds recorded in old rocks preserved in Greenland and South Africa. Single-cell life appeared when she was 11 (based on stromatolites found in South Africa) and bacteria developed as she entered her teenage years. Like so many teenagers, much of that period is still a bit of a blur, but as she progressed through her later teens and early twenties, she gained more self-confidence, began to settle down, and got on with living. She experimented with new processes and new forms of life, some of which she would carry through to later years, whilst others, although worth trying out at the time, would be disregarded.

The first organisms containing chlorophyll did not appear until she was 26. They breathed oxygen into her atmosphere and oceans – an episode that has sometimes been termed the Big Burp – which laid the foundation for life as we know it today. At the age of 31, the first nucleated cells developed. By the time Mother Earth was 39 (at the end of the Precambrian Era), multicellular organisms had started to diversify in Australia, Europe, and North America.

Almost everything that people recognize on Earth today, including all substantial animal life, is the product of just the last six years of her life. By the time she was 40, animals with hard parts (bones, teeth, etc.) had developed, as witnessed by the fossils discovered in China and Canada (the Burgess Shale). Fish appeared when she was 41, but the land surface was virtually lifeless until she was almost 42, after which mosses started to invade the hitherto bare continents. Within the next six months ferns appeared, there was an explosion of aquatic life, and insects had arrived on the scene. A year later immense forests of tree ferns covered her body, dragonflies with 3-foot wingspans took to the air, and amphibians and amniotes (egg-layers) roamed her surface. Life, at last, had truly broken free of the water. At the age of 44, she went through another one of her fads when she fell in love with reptiles and her pets included the dinosaurs. Within six months, the first known birds had taken flight, together with bees and beetles. The break-up of the last supercontinent was in progress. It was nearly a year later before she noted the arrival of flowering plants and the planet began to take on the appearance we see today.

Six months ago, dinosaurs went out of favour and she turned her attention to mammals, which largely replaced them in her affections. Primitive tools, found in Ethiopia, indicate that two and a half months ago "intelligent life" began to interfere with her landscape. About ten days ago, some man-like apes, living in Africa, turned into ape-like men. Last weekend, she began to shiver her way through the latest, but by no means the only

Table 1.1 The geological time scale. Each value (×1 million years) relates to the line below the value

Eon	Era	Period		Epoch	Date
Phanerozoic	Cenozoic	Quaternary		Holocene	0.01
		Tertiary	Neogene	Pleistocene	1.8
				Pliocene	5.3
				Miocene	23.7
			Palaeogene	Oligocene	33.7
				Eocene	54.8
				Palaeocene	65
	Mesozoic	Cretaceous		Late	99
				Early	144
		Jurassic		Late	159
				Middle	180
				Early	206
		Triassic		Late	227
				Middle	242
				Early	248
	Palaeozoic	Permian		Late	256
				Early	290
		Carboniferous		Late	323
				Early	354
		Devonian		Late	370
				Middle	391
				Early	417
		Silurian		Late	423
				Early	443
		Ordovician		Late	458
				Middle	470
				Early	490
		Cambrian		Late	501
				Middle	513
				Early	543
Precambrian	Proterozoic	Late			900
		Middle			1,600
		Early			2,600
	Archean	Late			3,000
		Middle			3,400
		Early			3,800
	Hadean				4,600

cold period in her life. Around four hours ago, a new, upstart species of animal, calling itself *Homo Sapiens*, took their first tentative steps in trying to take over the Earth. In the last hour, they invented agriculture and began to turn their back on a nomadic life style. It was only a quarter of an hour ago, that Moses and the Israelites crossed the Red Sea, and it is less that ten minutes since Jesus preached in the same area. The Industrial Revolution began less than two minutes ago, but in that brief time, out of the lady's 46 "years", we have managed to use up a substantial proportion of her resources – many of which she had taken a significant proportion of her life to produce. It is only in the last 10 seconds that we have begun to understand the nature of Mother Earth and the damage we are doing to her and her atmosphere. How many more seconds will it be before we start to treat her with the respect she deserves?

The geological time scale, presented in Table 1.1, will prove useful, not only in relation to the above but also with regard to Chapters 2 and 3.

Discussion point

Reflecting on the geological time scale, is there anything that takes you by surprise?

Is it the fact that we know so little about Mother Earth until she is well into middle age or that life, as we know it, is so recent?

Does it put our current concerns over climate change and the speed of change into context?

2
Dating rocks

2.1 Introduction

This chapter provides a brief overview of some of the methods used to date rocks and rock sequences. If you would like to find out more, you can find a comprehensive explanation of this topic in most basic geology textbooks. It does not include methods such as dendochronology – counting and comparing tree rings or varves – which are thin layers of alternating light and dark coloured layers of sediments that form glacial lakes. Both of these methods are generally only used to date fairly young rocks, in a geological sense.

Discussion point

Do you think the dating of rock should be an important part of geology and Earth Science?

The problem with almost any rock sequence is that, no matter how extensive it is, it can only cover a relatively small area of the globe. This means that trying to tie a sequence of rocks in one place to the rocks somewhere else can be difficult. Add to this the fact that the type of rock laid down in any particular place is dependent on the environment in which it was formed, or it may also have been folded and faulted, or altered by high pressures or heat – each of which adds a level to the complexity of unravelling the geological history of the rocks.

There are two broad methods used to date rocks. The first involves comparative processes in which the rocks and fossils are ordered in a sequence that indicates that they are either older or younger than each

Time Matters: Geology's Legacy to Scientific Thought, 1st edition. By Michael Leddra.
Published 2010 by Blackwell Publishing Ltd.

other. This is an easy task when fossils show progressive changes, but with rocks it is important to think about the conditions in which they were formed. If they are sedimentary rocks – rocks that are formed by the laying down of clays, sand, etc., or built up by organic activity such as reefs – you can often see how changes in the environment produced changes in these rocks. If they are igneous rocks – formed from molten rock either deep below or on the surface – you can see how they relate to, and maybe how they have affected, surrounding rocks. Each of these comparative methods provides a **relative** age, which does not have an implied date unless other factors are taken into account. This may vary from an understanding of how long it takes particular igneous rocks to cool down or typical rates of sedimentation based on observable data. It is only when you introduce such concepts as uniformitarianism that you can predict approximate dates to relatively dated sequences or when you have dates obtained from the second method, known as absolute dating.

Absolute dating allows a rock to be given a specific date or a range of dates in which it was formed, and is based on the measurement of processes such as radioactive decay.

This chapter will begin with a review of the development of the principles of stratigraphy, together with the origins, principles, and use of relative dating. This will be followed by an outline of radiometric dating.

2.2 The nature of stratigraphy and the principles of relative dating

In their book, *Unlocking the Stratigraphical Record: Advances in Modern Stratigraphy*, Peter Doyle and Matthew Bennett describe stratigraphy as "the key to understanding the earth, its minerals, structure and past life". They define it as "the study of rock units and the interpretation of rock successions as a series of events in the history of the earth". Although it may seem obvious today that rocks and the fossils they contain form in a sequential order, it was not until the 17th century that this was recognized and that consequently the rocks must contain an historical record of their formation.

Nicolaus Steno (Fig. 2.1), physician to Grand Duke Ferdinand II of Tuscany in 1669, whilst working in western Italy, was the first to recognize that rocks show a sequential change. Originally, he studied anatomy and then progressed to studying the origins of fossils, eventually publishing his ideas under the title *De solido intra solidum naturaliter contento*

Fig. 2.1 **Nicolaus Steno**

dissertationis prodromus (Prodromus to a dissertation on Solids naturally enclosed in Solids). This was eventually translated into English as *Prodromus of Nicolaus Steno's dissertation concerning a solid body enclosed by process of nature within a solid* and was published by the Royal Society in 1671. Although this was supposed to be a preliminary "discourse" (*prodromus*), Steno never wrote the final text. He attempted to explain how one solid body (a fossil) could be enclosed in another solid body (a rock). Steno observed that objects, which resembled parts of organisms, occurred in sequences of layered rock and decided that these were fossils rather than objects that grew in the ground, which were known as "*Glossopetrae*".

Steno recognized the Principle of Superposition in that "at the time when any given stratum was being formed, all the matter resting upon it was fluid; and therefore at the time when the lowest stratum was being formed, none of the upper strata existed". This marked the beginning of stratigraphy that is "the science of geological strata".

Put simply, superposition says that in a sequence of rocks the youngest rocks are at the top and the oldest are at the bottom (Fig. 2.2). This might seem a bit obvious but when the rocks have been tilted, possibly even vertically or overturned, this may not be so clear.

Background

The way that geologists determine what is the top and bottom of a particular layer of rock is by looking for "way-up structures". These are natural features related to the way in which the rocks or the fossils they contain reveal the order in which they were deposited. To illustrate this, I shall use a few examples:

1. Plants and trees grow vertically if they can, and you would expect to find their roots at the bottom and their trunks, branches, and leaves above the roots. Therefore, if you find plant or tree roots attached to a stem or trunk preserved in a rock, you can determine which are the top and bottom surfaces of the rock in which they are contained. Often the upper ground surface can be identified in which the plants or trees grew.
2. The upper and lower boundary surfaces of a layer of sedimentary rocks – called bedding planes – represent breaks in deposition. These may be due to simply a hiatus in sediment supply, or erosion, where some of the original sediments have been removed. Where sediments are deposited by water or wind, they generally show a graduation in particle size, with the larger particles forming the bottom of a layer of sediments. Particle size then decreases upwards through the layer, with the finest particles forming the top of the layer. This is referred to as a fining-upward sequence and provides a good way-up structure.
3. Following on from the example above, if the sediments are deposited in moving water or wind, they usually form some kind of ripple or dune. The internal structure of these comprises individual layers of fining-upward sequences. If deposition has stopped through erosion, the upper surface is truncated or chopped off. But how do we recognize which one is this surface? Ripples or dunes show a smooth-curved transition at the base that gradually becomes steeper until the curved lines are chopped off, usually leaving a relatively sharp, angular change with the bed above.

If you would like to find out more about this fascinating subject, I would recommend looking at any good sedimentology, stratigraphy, or general textbook.

Steno also developed the Principle of Original Horizontality, which meant that, "strata either perpendicular to the horizon or inclined towards it were at one time parallel to the horizon". He recognized that most sediments were laid down in water, which means that gravity causes them to settle out in layers parallel to the horizon. This means that in the majority of cases, if a sequence of rocks is not horizontal, they must have been disturbed (Fig. 2.3).

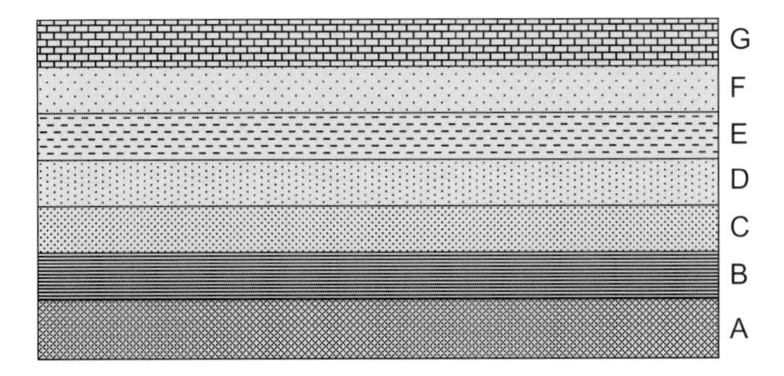

Fig. 2.2 **A diagram illustrating the Principle of Superposition, in which rock A is the oldest in the sequence and was therefore deposited first; rock G is the youngest and was the last to be laid down**

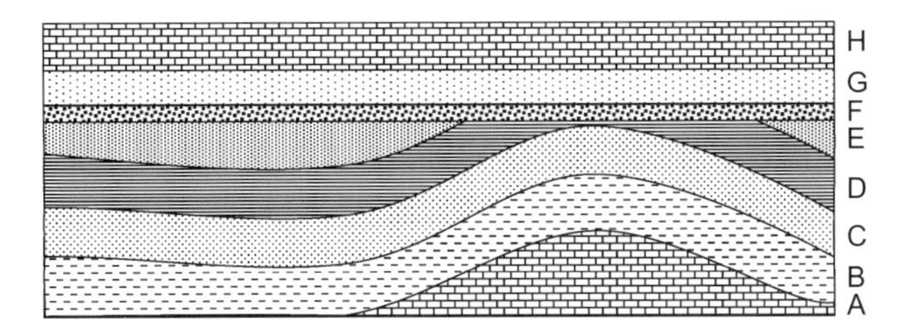

Fig. 2.3 **A diagram illustrating the Principle of Original Horizontality, which means that the rock units A to E must have been disturbed before F to H were deposited**

He also proposed the concept of Original Lateral Continuity. He recognized that sediments laid down in water are spread as laterally continuous sheets that are only limited by the edges of the basins or seas and the depositional conditions in which they have formed.

Superposition indicates that sedimentary rocks are laid down in sequential order, therefore if any rock cuts across the sequence or is out of alignment, this denotes that it is younger than the rocks in the sequence. The stacking of dunes or ripples in Fig. 2.3 above provides an illustration of this process.

Figure 2.4 represents an example of the Principle of Cross-cutting Relationships. This principle indicates that if you find a sequence in which the layers of rock are cut through by a different rock unit, the cross-cutting rock must be younger than the rocks through which it passes. Generally, however, most people think of cross-cutting relationships with regard to either igneous intrusions – where molten rock has been injected into or

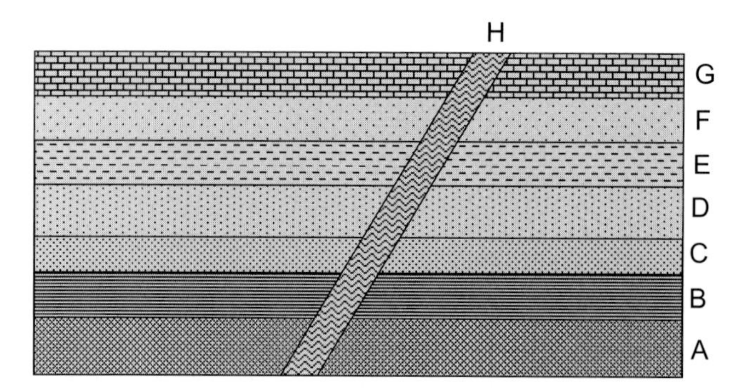

Fig. 2.4 **A diagram illustrating the Principle of Cross-cutting Relationships. The rock sequence is the same as that in Fig. 2.2, but there is an additional rock (H) that cuts across each of the other rocks and therefore has to be the youngest rock in the sequence**

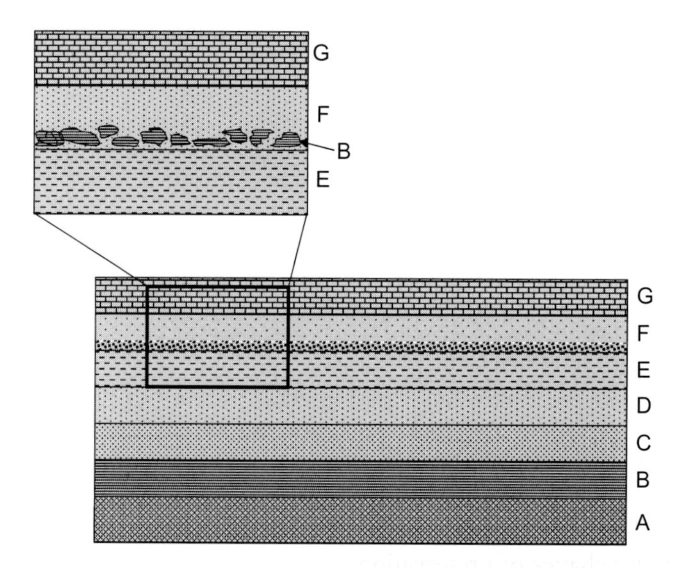

Fig. 2.5 **A diagram illustrating the Principle of Inclusions. Again the rock sequence is the same as that in Fig. 2.2 with the addition of pieces of rock in rock layer F. The expanded diagram shows that these are pieces of rock B which has to have existed before rock F was deposited, for them to have been included in that sequence**

across the rock sequence – or through faulting, where the rocks have fractured and been displaced relative to each other.

The Principle of Inclusions (Fig. 2.5) covers the presence of any particles of rock, fossils, or any other items that are preserved in a rock. This principle means that if any of the above are found in a rock, then they have to be contemporary with or older than the rock in which they are preserved. Sandstone, for instance, is composed of sand grains that have been eroded and transported to the place of deposition. The sand grains must therefore be older than the sandstone in which they are found. One of the implications of this is that any date attached to sandstone denotes the age of the sandstone, not the age of the sand grains from which it comprises. Equally, if there are fragments of other rocks included in a layer of rock, the fragments must be older than the rock layer in which they are found, as they have already gone through the processes of formation/deposition, lithification, erosion, transportation, and subsequent re-deposition.

The presence of fossils follows the same idea as the Principle of Inclusions. The fossils have to be contemporaneous with or older than the rocks in which they have been found. Generally, you would look for evidence that they are in "life position", i.e. there is clear evidence that they were alive in the position in which they have been preserved. If not, there is usually evidence that they have been moved from their life position and been transported to their present position.

2.3 Biostratigraphy

When you consider the variety of living plants and animals on Earth at a single "moment" in geological time, you quickly realize that this is an immense subject. This variety includes types (genera and species) that have become specialized to live in a wide range of different environments, from equatorial forests to the snow and ice of the polar regions, from the top of mountains to the depths of the oceans. Most sedimentary rocks that are preserved in the rock record were deposited in the oceans, rivers, and deltas or deserts. This therefore means that most fossils that are found are usually associated with these environments. Of these, those that lived in the oceans have a better chance of preservation.

The use of the term "chance" refers to the likelihood of something being preserved as a fossil. Firstly, it generally has to be composed of something that can be preserved. Most fossils preserve evidence of bones and shells or other hard parts, as they are more resistant to being eaten or decomposure. The organisms generally have to be living in an environment that either allows preservation, or they have to have been transported to an environment in which they can be preserved. This is usually underwater, where decomposition or the movement of the sediments or the animals, plants, etc. is reduced. The organism then has to be buried by sediments to further reduce its chances of being eaten. Burial usually slows down the rate of decomposure. Finally, the rocks themselves have to be preserved. If they are weathered or eroded (broken down and moved), the chances of the preservation of the fossils they contain are significantly reduced. Consequently, it is interesting to think about how many of today's living things could be preserved as fossils. Given the limitations outlined above, which animals and plants, etc. will provide future geologists with clues about the rocks and environments in which they lived, and how representative of life in general will they be?

The way in which fossils have been viewed and used in the past is considered in Chapter 7.

Background

Biostratigraphy refers to the dating and correlating of rocks by means of fossils. This involves the study of sequences of fossils that have three characteristics:

1. Fossils that evolve through time so that the fossils, and therefore the rocks containing them, can be placed in sequential order;
2. Fossils that have a widespread habitat (areas and environmental conditions in which they live), so that rocks deposited in a number of different environments can be tied together in time;
3. Fossils that have very specific habitats and are therefore good indicators of environmental change.

Difficulties arise because:

1. Rates of evolution vary greatly.
2. Extinctions occur fairly frequently.
3. There is a natural bias in the types of organisms that are generally preserved as fossils.

Modern biostratigraphy uses microfossils (microscopic-sized fossils) such as Foraminifera, as these are usually found in large numbers in many types of marine sedimentary rocks. They are also more likely to be found in smaller samples – such as cores from boreholes – than macrofossils, which are fossils visible to the naked eye. For the same reasons, plant spores and pollen are often used when studying terrestrial rocks.

Discussion point

Alan Cutter in his book about Steno, *The Seashell on the Mountain Top*, makes the point that although "understanding that a stack of objects gets piled up from bottom to top requires no great intellectual leap", the realization "that layers of bedrock contained a narrative, that it made sense to speak of one rock being 'older' than another, was Steno's critical breakthrough". He also includes the following comment: "It is an amazing fact of the history of science that before Steno few European writers had thought this fundamental observation worth mentioning".

Although Steno established his ideas in the 17th century, Cutter points out that it was not until the 18th century that his ideas were recognized as being important. Just as he was disappearing from the history of science, to be forgotten for over a century, "the British scientist James Hutton supposedly invented modern geology and opened up deep, geological time for exploration, where they 'rediscovered'". It was only after this that Steno was belatedly recognized as one of geology's founding fathers, and even then, his apparent abandonment of science, to follow his faith, meant that he was viewed as a lesser scientist.

Using the above principles – which are still used in the field today – natural philosophers were able to begin constructing theories and classifications for the rock sequences they studied.

Able Moro (1687–1740), a Roman Catholic priest and naturalist, studied the volcanic eruptions of Santorini. He concluded that all crystalline mountains were formed by the violent actions of subterranean forces that he called "primary mountains". He thought that other mountains, which were made up of layered rocks, were formed later on and so he called these "secondary mountains", as they consisted of eroded volcanic material and other sediments.

During the latter half of the 18th century, the move away from studying collections of rocks and fossils in museums to undertaking fieldwork continued. This change from looking at collections assembled by other people to making your own observations was deemed essential. As Rudwick writes in his book *Bursting the Limits of Time: The Reconstruction of Geohistory in the Age of Revolution*, it showed that you had to have seen something with your own eyes, and studied it, before you could make a pronouncement about it. Fieldwork therefore became a necessary step for anyone who wanted to "establish any credibility or authority" for their work. Freeman reiterates this when he includes a definition of geologists used by William Thomas Brande, the famous chemist, at the start of his lectures on geology in the early part of the 19th century:

> Persons have been called Geologists who, gifted with prolific imaginations have indulged in fanciful speculation concerning a former order of things … Others, by careful, diligent, and extended observations of the present

Fig. 2.6 **Johann Gottlob Lehmann**

> state of the earth's surface, have endeavoured in the path of induction, to trace the nature of the agents which have once been active, to ascertain how they are now operating, and to anticipate the results of their continuance … These are really Geologists, and their aim is not to imagine or suppose, but to discover …

In 1756, Johann Gottlob Lehmann (1719–1767) (Fig. 2.6) – a physician, copper producer, and Professor of Chemistry in St Petersburg – and in 1761 George Christian Fuchsel (1722–1773) – a German geologist who published one of the world's first geological maps – separately applied the principle of superposition to two areas in Germany.

Lehmann based his classification on the physical characteristics of the rocks, which he thought could be divided into three groups:

1. Primitive mountains formed at the time of creation;
2. Secondary mountains, which were layered rocks lying against the primitive rocks;
3. Tertiary rocks, which included the products of volcanoes.

He thought that he could extend the use of his classification to all the rocks around the world. He also thought that this would help miners predict where particular rocks could be found at the Earth's surface and allow them to predict where they would occur underground as well. He was probably one of the first people to use his own observations and classification to try to understand how different types of rocks were formed; he regarded himself as a historian of geology.

In 1760, Giovanni Arduino (1714–1795) – a Mining Inspector and Professor of Mining in Venice, who is known as the father of Italian geology – devised a four-fold classification:

1. Primary rocks, which were the crystalline rocks with metallic ores;
2. Secondary rocks, which were hard, stratified rocks without ores but with fossils;
3. Tertiary rocks, which were weakly consolidated stratified rocks usually containing numerous shells of marine origin. This division also included volcanic rocks.
4. Alluvium, which consists of material washed down from mountains.

Peter Simon Pallas (1741–1811), born in Berlin, was a German zoologist and botanist that worked in Holland and Britain before going to Russia. In 1761, he published his observations and interpretations of corals and sponges. In 1767, he moved to the St Petersburg Academy of Sciences and in 1768, he became the Chair of Natural History. He led a number of expeditions of naturalists and astronomers, commissioned by Catherine II, across Russia. Pallas supported the views of Able Moro that the Earth had an igneous core and recognized a three-fold sequence in the Ural Mountains in which he considered that the granite cores were formed at the time of creation and later rocks lay against them at steep angles.

The three-fold system was by then widely recognized throughout Europe. People were beginning to think that this system represented the time span of the Earth's history and started to wonder if they represented the same time spans in different locations. In doing so, it is interesting to note that Cuvier (see below) considered that Pallas had created a completely new geology.

Abraham Gottlob Werner (1749–1817) (Fig. 2.7), born in Wehrau in southeastern Germany, studied law and mining at Freiburg and Leipzig but eventually dropped law to concentrate on mining and mineralogy. He was appointed as an Inspector and Teacher of Mining and Mineralogy at the Freiburg Mining Academy in 1775, and was the curator of the academy's mineral collection. A year earlier he had published what is considered the first textbook on descriptive mineralogy, entitled *Vonden ausserlichen Kennzeichen der Fossilien,* which has been variously translated as *On the External Characters of Fossils* or *On the External Characters of Minerals.*

Fig. 2.7 **Abraham Gottlob Werner**

Background

The word fossil comes from the Latin word *fossilis*, meaning "that which is obtained by digging". The term was originally used for any "curious or valuable" rock or mineral, as well as what we would now call fossils. Indeed, until the latter part of the 18th century and the early part of the 19th century, people regarded rocks and minerals as "natural fossils", whilst those that we would regard as fossils today were called "accidental fossils".

Werner was particularly interested in the systematic identification and classification of minerals. It is interesting to note that the Mining Academy had been set up only ten years earlier in 1765, to "provide practical training in assaying, engineering and mineral science", and had "developed into one of the premier schools of mineralogy in Europe" in a relatively short period of time.

Werner thought that individual rock units were universally distributed. It was subsequently realized that one of the major problems with his ideas was that, although they worked for the rocks in Saxony where he lived, there were significant problems when this classification and its interpretation were applied elsewhere for the formation of the different types of rocks.

Werner is regarded as the person who established mineralogy as a distinctive science by devising his own systematic classification. In 1787, he published his *Short classification and description of different rocks*. He gained a reputation for his personality and teaching style and from his imaginative approach to the subject that helped turn the Freiburg Academy into one of the most important centres for science in Europe.

He introduced the term "Geognosy", which meant "knowledge of the Earth", to define a science that was based on the recognition of the order, position, and the relation of the layers that form the surface of the Earth. Together with his comprehensive classification for minerals, Werner emphasized the orderly sequence of rocks, which he divided into different formations, devising a four-, and later a five-fold classification:

1. *Urgebirge* (Primitive rocks), which he thought were formed from chemical precipitates. In this group he included granite, gneiss, schist, slate, and other crystalline rocks. He considered these to be universal formations that were the first ones laid down when the ocean still covered the entire globe up to the top of the highest mountains.
2. *Uebergangsgebirge* (Transitional rocks) were slate, schist, greywacke (mudstone and muddy sandstone), and some limestones with occasional fossils. These were also universal formations, which he thought were the first rocks to be deposited as the universal ocean started to recede.
3. *Flotzebirge* (Stratified or Secondary rocks), which included sandstones, evaporates, limestones, and basalt formed from the products of erosion as the oceans continued to shrink. He considered these to be the first, non-universally deposited rocks. In other words, they had a limited spatial distribution.
4. *Aufgeschwemmte gebirge* (Washed-up rock, Alluvium). This included gravels, unconsolidated sands, and clays that formed as the level of water continued to drop and an increasing proportion of the land surface was exposed to erosion. He considered these to be the products of disintegrated primitive rock.
5. *Vulkaniche gesteine* (Volcanic rocks), which included lava flows and pyroclastic rocks associated with volcanic vents. Werner thought that the localized, subsurface burning of coal seams formed these.

He was also one of the first people to formulate a theory for the formation and history of the Earth (see Neptunism in Chapter 4).

Discussion point

Werner recognized that, because of the above sequence, each rock type occurred at its own horizon and therefore represented its own unique time of origin. In other words, the geological succession showed a progressive sequence of events (see Hutton and Lyell in Chapter 5).

Georges Cuvier (1769 1832) (Fig. 2.8) was born in 1769 in Montbeliard in the Jura Mountains. In 1795, he became an assistant and later a professor of animal anatomy at the National Museum of Natural History in Paris. He was also appointed to the position of Inspector General of Public Education and became a State Councillor under Napoleon. According to Douglas Palmer in his book, *Earth Time: Exploring the Deep Past from Victorian England to the Grand Canyon*, Cuvier wrote, "would it not also be glorious for man to burst the limit of time, and, by a few observations, to ascertain the history of the world, and the series of events which preceded the birth of the human race?"

Fig. 2.8 **Georges Cuvier**

Although Cuvier is most famous for his work on fossils and animal anatomy, he spent a great deal of time studying the rocks of the Paris Basin, during which he recognized that the rocks followed a particular order and that each one contained specific groups of fossils (Chapter 4).

As the rock sequence in the Paris Basin was incomplete, Cuvier was unable to find the links between different fossils. He therefore concentrated his investigations on the breaks between the different rock units, as the variation in the types of rocks and the fossils they contained showed significant changes. He thought that each of these changes represented a break caused by a large worldwide flood that destroyed many of the organisms that existed at the time (Chapters 5 and 6).

William Smith (1769–1839) (Fig. 2.9), the son a blacksmith, was born in Churchill, Oxfordshire in 1769. He became an engineer and surveyor and worked on roads, quarries, mines, and canals all over the country. This work allowed him to study a huge array of rock sequences and the fossils they contained. This experience enabled him to distinguish between similar types of rocks (lithologies) of different ages. He also recognized that the fossils contained in one sequence of rocks were not repeated exactly in the rocks above and below it. This meant that the fossil assemblage – that is, the different types of fossils in a particular layer or sequence of rocks – could be used to date the rocks relative to one another, a process that was subsequently known as the Law of Faunal Succession. It also enabled him

Fig. 2.9 **William Smith**

to predict where particular sequences of rocks would occur and allowed him to link together rocks of similar ages in different places.

In 1793, Smith began work on the construction of a canal near Bath: he built up first-hand knowledge of the strata around the city and its order of superposition. He used two criteria for his work:

1. That the lithological units were distinctive;
2. That they contained a distinctive fossil assemblage.

Often rock sequences do not contain just a single, diagnostic fossil. They usually contain a suite of fossils, each of which may have evolved at a different rate. This fossil assemblage is unique to that particular rock sequence. When Smith moved to London, he set out his fossil collection, not in the usual manner of grouping the same types together, but by placing them in the stratigraphic order – indeed, Smith coined the term "stratigraphy". (See Rudwick for a particularly good review of Smith's work.)

Smith found that particular groups of fossils were always associated with specific rock sequences and that these sequences of rocks were always found in the same order. This allowed him to do two things: firstly, he could use this to predict where a particular type or sequence of rocks would exist; and secondly, it allowed him to draw the first geological map of Great Britain, which he published in 1815. This was titled *Geological Map of England and Wales and Part of Scotland; exhibiting the Collieries and Mines; the Marshes and Fen Lands originally Over-flowed by the Sea; and the Varieties of Soil according to the Variations in the Sub Strata; illustrated by the Most Descriptive Names.* He also published 21 separate maps, the *Geological Atlas of England and Wales*, which was the first attempt to map the distribution of geological sequences over the whole country.

In his book about the life of William Smith, titled *The Map that Changed the World*, Simon Winchester provides a very good appraisal of his work, his life, and its importance to the development of modern geology. The dust cover gives an insight into the difficulties he faced as an ordinary person in a world where money and status were so important. As Winchester and Rudwick describe, Smith was a self-taught man who, through his engineering and geological work, enabled him to investigate, at first-hand, the relationships between rock units and the fossils they contained. This allowed him to construct a map that, although other maps already existed, presented geological strata in a unique, almost three-dimensional way – a technique that has largely continued to this day. As Winchester says:

> Smith's was a remarkable achievement and all the more astonishing for having been completed single-handedly and without financial or professional support. Shatteringly, such heroic and painstaking work exacted a terrible price: imprisoned for debt, Smith was turned out of his home; the

> work was plagiarized; the scientific establishment turned its back on his troubles; and Smith's wife was diagnosed insane and he himself fell ill. It was not until 1829 that, in a fairytale twist of fate, Smith returned to London in triumph, to be hailed as a genius.

Doyle and Bennett point out that Smith named many of the Mesozoic rock units that are so well-known today, such as Chalk, Greensand, Blue Marl, Forest Marble and Clay, Millstone, Magnesian Limestone, Coal Measures, and Derbyshire Limestone. They make the point that "although these were some of the earliest lithostratigraphical units to be delineated, the fact that many form the basis of modern, formally defined, units shows the basic soundness and objectivity of Smith's criteria for subdivision".

It is important to note that the development of geological thinking in Britain was significantly different to that in the rest of Europe. At the time (the mid-1830s), the UK was dominated by "gentlemanly specialists" who used the Geological Society of London – which they founded on the 13 November 1807 – to help to define themselves as an "elite group". Amateurs were regarded as mere fact-gatherers. It is for this reason that Smith was almost totally ignored at the time.

Winchester points out that the Geological Society had been set up as:

> a social and dining club for the purpose of making geologists acquainted with each other, of stimulating their zeal, of inducing them to adopt one nomenclature, of facilitating the communication of new facts, and of contributing to the advancement of geological sciences.

Smith was considered to be "unpolished and ill-educated", with a "common accent" and therefore was not allowed to become a member of the society. He was dependent for his living on the practical application of geology. (See Chapter 7 for further examples of the attitude of the Geological Society in the 19th century.)

Winchester also explains that in the early spring of 1808, a "band of distinguished Londoners arrived at Smith's front door in Buckingham Street". They had come to see his fossil collection and map. The group was led by George Bellas Greenough (Fig. 2.10). After the visit, Greenough and James Hall decided to create their own map that would be "the definitive and official geological map of the country". The only way they could really do this was by copying Smith's map and adding their own information to it. This was a problem, as Greenough in particular did not agree with the "European ways" of doing geology that Smith had adopted and he therefore had great difficulty adding his and the society's information to the map. It took longer than expected to produce and Smith managed to publish his first edition in 1815, whereas Greenough's map did not appear until 1819. However, Greenough's map was sold at a lower price to

Fig. 2.10 **George Bellas Greenough**

Fig. 2.11 **The building in which William Smith lived whilst in Scarborough, North Yorkshire (left and insert), and the Scarborough City Museum, the Rotunda, designed by him (right)**

undercut Smith. It was this that helped ruin Smith and put him in the debtors' prison.

He was forced to sell everything – his fossils went to the British Museum, whose mineralogist, Charles Konig, did not even unpack them. On his release, he moved to Scarborough, North Yorkshire, where he helped to set up the Scarborough City Museum, the Rotunda (Fig. 2.11), to hold and display fossils in their stratigraphic order.

It was not until 1831 that Smith was finally given the recognition he deserved, when he was presented with the first-ever Wollaston Medal, which is regarded as the "oscar of the world of rocks". In his address at the meeting where Smith was presented with his medal, Adam Sedgwick (see the origins of the geological time scale in Chapter 3) referred to Smith, who by then was 62 years old, as the "Father of English geology". In 1865, the Geological Society eventually went one stage further and made an addition to the title of Greenough's map, *A Geological Map of England and Wales, by G. B. Greenough Esq., FRS (on the basis of the original map of Wm. Smith, 1815)*. Finally, the source of the data had been acknowledged and Smith's pioneering work was officially recognized. If you want to see how the establishment can work against someone who they think is not their equal, Winchester's is a very good book to start with. Smith's story and the behaviour of some of the country's leading scientists at the time provide a lesson, which all those who are now in positions of power, authority, and responsibility should learn from. It is particularly worth reading chapters 13 and 17 of Winchester's book, which gives a good insight into the geological world in England at that time.

Smith, Buckland, William Conybeare, and William Philip's work showed, as Rudwick puts it, that "the British sequence of formations turned out to be exceptionally straightforward and undisturbed; it had the potential to serve as a standard of comparison for international and even global correlation". In fact, as Rudwick explains, Conybeare and Philips book *Outlines of the Geology of England and Wales*, published in 1822, became an "indispensible handbook" that helped this methodology become "the envy of Continental geologists and gave them an incentive to correlate their own formations with what was soon being treated in practice as an international standard of comparison".

Smith's influence did not end there; Cuvier and Brongniart extended his use of fossils to establish particular geological formations to indicate that they also said something about the environment in which they lived.

Discussion point

His humble origins meant that he would probably not have been taken seriously by the "professionals" of the time, even though – or probably because – he was a "practising geologist". In his book *Victorians and the Prehistoric: Tracks to a Lost World*, Michael Freeman describes Smith as being "among the prototype field geologists and perhaps the last to come entirely from a non-academic background". At the time, geology was for the well-to-do, those with money and status who regarded themselves as thinkers rather than doers. The following generation of geologists began to change this attitude, hence trying to right the injustice done to Smith.

Is there still a similar "divide" in geology in particular, and science in general?

It is important to note that in the early 19th century scientists knew that the Earth was ancient, but they had not devised a scheme for ordering the different events into an actual history. They had begun to realize that life, for instance, showed a continuous line of development and that species could become extinct. The establishment of a history, i.e. the geological time scale, which happened over a relatively short period of time, was probably one of the greatest contributions geology has ever made to the development of science.

Following the work of Smith and Cuvier, it became possible to define major units of sedimentary rocks, which were known as the Geologic Systems. These grew without a plan through the efforts of numerous geologists working largely independently (Chapter 3).

Alexander von Humboldt (1769–1859) (Fig. 2.12), a German naturalist and explorer who was a catastrophist and a student of Werner, was the first person to realize that the Earth's present environment is a product of a long series of changes.

He spent a great deal of time exploring South America for the Spanish Government. In 1845, he published his five volumes of *Cosmos: A sketch of the Physical Description of the Universe.* As Gerard Helferich records in

Fig. 2.12 **Alexander von Humboldt**

his book *Humboldt's Cosmos: Alexander von Humboldt and the Latin American Journey that Changed the Way we See the World*, this "massive work attempted to outline all knowledge about the physical sciences in a way that would reveal to the intelligent lay reader the order underlying the universe's apparent chaos". Helferich points out that one of the reasons for publishing such a weighty publication – which was based on a series of lectures that Humboldt gave in Berlin between 1827 and 1828 – was to "counter the unscientific, romantic speculations of the German 'natural philosophe'". He created the science of plant, physical and political geography, oceanography, and climatology and was "instrumental in focusing scientists' attention on the need for accurate, systematic data collection".

Discussion point

Although this relates to Chapter 5, it is worth making the point here that catastrophists had a major influence on geological thinking at the time. They were very important in the development of modern stratigraphic thinking, as their ideas were usually based on careful observation and good reasoning. They proposed that the Earth's history had a direction whereas, as we shall see in Chapter 5, the uniformitarianists tried to replace directionalism with cyclic or steady-state models.

Gould has argued it in his book, *Time's Arrow, Time's Cycle*, in which the discovery of deep time (a long geological time scale) combined the insights of theologians, archaeologists, historians, and linguists, as well as geologists.

2.4 Radiometric dating

Although relative dating is valuable, particularly with regard to studying rock sequences in the field, it does not provide an absolute age. All that can be said is that one layer or group of rocks is younger or older than another. As can be seen from William Smith (above), this enabled him to identify repeated sequences, often by the fossils they contained, that meant he could place disparate rock layers into an overall sequence.

To be able assign a specific, accurate age to a layer of rock, it is important to find something that can be accurately dated. Chapter 1 introduced the concept of atomic (radiometric) dating. The following represents a brief overview of its principles.

This is one of the main methods used for absolute dating and is based on the principle that radioactive particles decay through unstable atoms

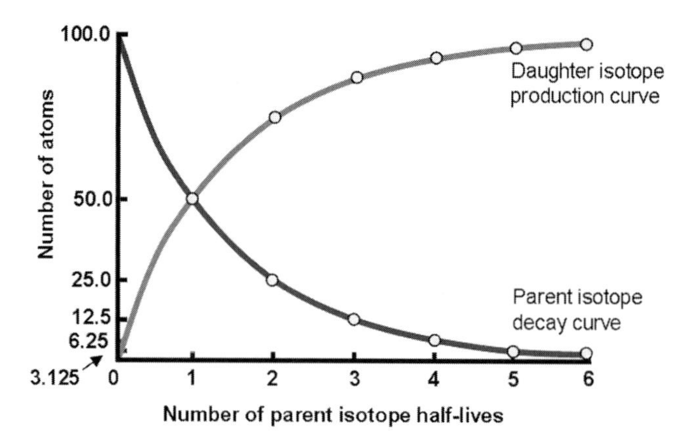

Fig. 2.13 **Radioactive decay and the production of daughter isotopes**

spontaneously changing to a lower energy, stable state by radioactive emission. As they do so, they release either alpha or beta particles or capture electrons. The original isotopes are referred to as the "parent" isotopes and the decay products are termed "daughter" isotopes (Fig. 2.13). If you wish to find out more about this, you can turn to most chemistry, physics, or geology textbooks. Again, the subject area is huge, so the following is a very brief overview.

Each radioactive nuclide has a particular mode and rate of decay that is independent of external conditions. Radioactive decay is known as the "half-life" and follows a geometric equation; this means that over a specific increment of time, the radioactive particle will lose half of its radioactivity. Over the next increment it will lose half again, so that over a period of two half-lives only a quarter of the radioactive material is left. Half-lives vary from microseconds to trillions of years. If a half-life is too short, there may not be enough of the parent isotopes left to count; equally, if the half-life is too long, there may be too few daughter isotopes to identify.

Some isotopes are vulnerable to natural processes that can allow isotopic material to be either gained or lost. For dating, the best nuclides are those whose half-life is roughly equal to the time span you are looking at.

Geologists tend to use two kinds of materials:

1. Long-lived ones, with half-lives from several millions of years to hundreds of millions of years;
2. Short-lived ones, with half-lives of tens of million of years.

As a radioactive material decays, it gradually converts to its stable daughter element. This means that if you know the half-life, you can use the proportionality between the radioactive element and the daughter element to determine how much radioactive material was originally present and over what period it has decayed. Usually dates are recorded for different isotopes in the same rock, which means that the dates can be cross-checked.

The four most widely used isotopes are as follows:

2.4.1 Potassium

This is the seventh most common element in the Earth, of which only 0.4% is the radiometric isotope ^{40}K. Of this, only 11% decays to ^{40}Ar (argon) and the rest converts to ^{40}C (carbon), which is non-radioactive and is therefore of no use for dating rocks. ^{40}K is particularly good for dating intrusive igneous rocks that naturally contain potassium-based minerals, such as mica and zircons. With a half-life of 1.3 billion years, it is especially useful for dating rocks between 100,000 and 4.6 billion years old. It is not commonly used to date igneous and high-grade metamorphic rocks that have been subject to very high temperatures, as this may have caused the minerals to release some of the argon gas. If the rock is heated to more than 200 °C or buried to a depth greater than 5 km, then some of the argon can be lost.

2.4.2 Rubidium

^{87}Rb (rubidium) converts to ^{87}St (strontium), but this isotope is not very common. However, if there is any original strontium around, it always contains ^{86}St, which is non-radioactive and therefore considered to be stable. It is then possible to compare the amount of ^{86}St present in a rock that has remained the same since the rock was formed, with the amount of ^{87}St that has formed as the original ^{87}Rb decays. Dating can therefore be achieved by checking the proportions of the two different types of strontium. This method is particularly good for dating metamorphic rocks where the rocks have been changed by heat and/or pressure. It is also useful for determining the ages of plutonic igneous rock, because ^{87}Rb has a half-life of 47 billion years and so is especially good for dating very old rocks, i.e. between 10 million and 4.6 billion years old. However, it is important to remember that these rocks have often had a very tortuous history.

2.4.3 Uranium

Bertram Boltwood first proposed this method of dating rocks in 1907 (Chapter 8). The two uranium (U) isotopes that are commonly used are

^{235}U and ^{238}U, which occur in the ratio 138 : 1. As ^{238}U has a half-life of 4.5 billion years and converts to ^{206}Pb (lead) and ^{235}U, which converts to ^{207}Pb, which has a half-life of 713 million years, the proportions of both can be used to cross-check the results. Both are found in zircons together with ^{235}Th (thorium), which has a half-life of 14.1 billion years and converts to ^{208}Pb. Uranium is a common trace element (i.e. it exists in small amounts) in many rocks. If you find ^{204}Pb, which is non-radioactive, present in the rock, you can use this to determine how much – if any – of the ^{207}Pb was originally formed at the same time and is therefore not the result of radioactive decay.

2.4.4 Carbon

^{14}C, which has a half life of is 5,730 years, is continuously generated 15 km above the Earth's surface by cosmic rays hitting ^{14}N (nitrogen) atoms. Both ^{14}C and non-radioactive ^{12}C react with O (oxygen) to form CO_2 (carbon dioxide), which living cells incorporate into their organic structures. ^{14}C converts to ^{14}N, but as long as a cell is alive, it will continue to take up both carbon isotopes in equal proportions. This means that organic samples can be dated by comparing the proportions of the two carbon isotopes. As ^{14}C has a relatively short half-life, it is best used to date samples between 100 to 70,000 years old.

2.4.5 Mass spectrometer

Radiometric dating is undertaken using a mass spectrometer. This produces a beam of electrically charged atoms from the rock or mineral sample. The chemical composition of rocks and minerals can be determined using this type of equipment, because each chemical has a different mass. Samples are vaporized to break them down into their chemical constituents, before the vapour is ionized and fired by a strong electrical field into a magnetic field. As the ions pass through the magnetic field, they are deflected by an amount that is dependent on their mass-to-charge ratio: the lighter ions are deflected the most and the heavier ones the least. The degree of deflection is recorded by a series of detectors that allows the chemical composition to be determined together with the proportion of each mineral within the rock. It can also be used to compare the relative concentrations of the different minerals. This method of dating rocks is commonly used for igneous rocks, where it produces dates with an accuracy of $\pm$ 0.2 to 2.0%.

Background

Various assumptions are made when determining radioactive ages and if there are any indications that any of these are incorrect, their effects have to be taken into account:

1. The system is closed, i.e. nothing is lost or added to the rock or minerals following their formation.
2. Neither the parent nor daughter atoms were added or removed from the system, other than by radioactive decay.
3. No daughter atoms were present in the system when the rocks or minerals were formed.
4. During cooling, parents are separated from existing daughters.

Few rocks conform to this because they contain original daughters as well as radiometric ones that, if they are not identified, will give an age greater than it should be.

Difficulties arise in determining the age of samples due to:

1. The partial loss of daughter atoms produces dates that are too young.
2. The loss of any parent atoms produces dates that are too old.
3. There may be errors in the laboratory analysis.
4. There may be variations among samples.
5. You have to assume that the radioactive conditions were the same in the past as they are today.
6. Weathering can cause a loss of isotopic materials.
7. Reheating can "reset" an atomic clock.
8. You assume that the half-life of an element is constant.
9. You assume that the speed of light is constant.

The scientists and laboratories involved in radiometric dating take great care to make sure that they cover the above and test multiple samples from which they determine the ages of the rocks.

Further reading

It is worth reading the section on radiometric dating (Chapter 7) of Eugene Scott's book *Evolution versus. Creationism: An Introduction*, which presents creationist and scientific arguments on the reliability of radiometric dating. I would also recommend chapter 6 of Paul Garner's book *The New Creationism: Building Scientific Theories on a Biblical Foundation* and Mark Isaak's chapter on geochronology in *The Counter-Creationism Handbook*. Garner reports on the eight-year study (that began in 1997) by a group of seven geologists, geophysicists, and physicists, funded by the Institute for Creation Research (ICR) and the Creation Research Council (CRS). This study, known as RATE (Radioisotopes and the Age of The Earth), looked at evidence for "accelerated radiometric decay". Garner's book should also be compared to the extremely good and balanced explanation provided by Isaak's and also with G. Brent Dalrymple's chapter *The Ages of the Earth, Solar System, Galaxy, and Universe,* in Petto and Godfrey's *Scientists Confront Creationism: Intelligent Design and Beyond.*

2.5 Dating by fission tracks

Alpha particles emitted by uranium isotopes in minerals are heavy, move fast, and damage the structures of crystals as they travel. These appear as linear tracks once the mineral has been etched with hydrofluoric acid in the laboratory. As nuclear fission occurs at a constant rate, the age of a sample can be determined by counting the number of tracks present. This technique can be used to date samples from 50,000 to billions of years old. It is most useful when dating rocks between 40,000 and one million years old, and therefore covers a gap not usually covered by other methods.

Background

If the mineral is subject to high temperatures, the tracks can heal and there will be fewer tracks than should have formed over the lifetime of the mineral. The resultant age calculation may therefore be lowered.

2.6 Magnetism

In 1958, Walter Elsasser (1904–1991) and Edward Crisp Bullard (1907–1980) proposed that the Earth's magnetic field is generated by convection currents in the its liquid outer core, which acts as a self-exciting dynamo. Patrick Blackett (1897–1974) won a Noble Prize in 1948 for proposing that any revolving body generates a magnetic field; he had been working on magnetic minerals. This theory was subsequently rejected, but rivalry between supporters of the different ideas led to a rush to record as much magnetic data as possible. (See Chapter 5 for Edmund Halley's hypothesis on the generation of the Earth's magnetic field.)

Fossil magnetism was supposed to have been first recognized by an Australian graduate when he looked at stones he had recovered from a bonfire in the outback; the stones had come from a cooking site dated at 30,000 BC. He noticed that the magnetic field held in the rocks had the exact opposite direction to that of today, i.e. the magnetic North and South poles had swapped over.

During the Second World War, extremely sensitive airborne magnetic equipment was developed to look for German submarines. This technology was later adapted for use by oceanographers in equipment that could be towed behind ships. When they started to look at the sea floor, they discovered that it had an amazing series of magnetic anomalies: these consisted of bands of positive and negative stripes that continued for hundreds of miles on either side of known oceanic ridges. It was realized that the bands on either side of the ocean ridges showed the same patterns of magnetic variations – generally, they were almost mirror images of each other (Fig. 8.12).

These reversals indicate periods when the Earth's magnetic poles swap positions, i.e. the magnetic North Pole swaps its polarity with the magnetic South Pole. These magnetic changes have two different components: a longer-term change known as secular variations, polarity epochs, or chrons that occurs about every 500,000 years; and shorter variations, known as magnetic events, polarity events, or subchrons that occur at periods of between 1,000 and 100,000 years with an average of around 4,000 years. It appears that during a reversal event, magnetic intensity decreases and both declination (the compass bearing of the magnetic pole) and inclination (the angle of the magnetic field compared to the horizontal) becomes irregular and may even be multipolar.

In 1963, Fredrick Vine and Drummond Matthews (Chapter 8), who were English marine geologists and geophysicists, and Lawrence Morley, a Canadian geophysicist, independently came up with the theory of sea-floor spreading to explain the above. They showed that when the magnetic data was tied in with radiometric dating of the rocks for the last 7 million years, the polarity of the Earth's magnetic field has frequently reversed over a long period of geological time. A Dutch geologist, Martin Rutten, produced the first published data on the subject in 1959 and Cox published the first magnetic history that covered the last 5 million years in 1969.

Using deep-sea cores, the Earth's magnetic history has since been extended back to between 100 and 200 million years and it has been possible to use magnetic data to date rocks as far back as 3,500 million years. One of the basic problems with magnetic data is that the further back in time you go, the more likely it is that the rocks have been subsequently heated above their Curie point (the temperature at which magnetic minerals loose their magnetism) and their original magnetism has been lost.

Magnetism in rocks is usually generated in two ways; thermal remnant magnetism (TRM) or depositional remnant magnetism (DRM).

2.6.1 *Thermal remnant magnetism*

When a liquid magma cools to form an igneous rock, certain minerals, for instance magnetite, acquire magnetism at a critical temperature known as the Curie point. The orientation of the magnetic minerals is determined by the Earth's magnetic field until crystallization occurs. From then on, the orientation is effectively frozen unless the rock is reheated to a temperature above the Curie point. This generally occurs at a temperature between 500 and 600 °C, which exists at a depth of between 20 and 30 km below the Earth's surface. This means that any magnetic minerals that exist below that depth cannot be permanently magnetized. It has been noted that most magnetism occurs in the top 0.5 km of the basalts that form the oceanic crust.

Background

Experiments have shown that some Fe (iron)-Ti (titanium) oxides that have different Curie points can become magnetized with a reversed polarity. Examples of reversed magnetism in Fe-Ti rocks have been found in some young lava flows, but it does not appear to be a major cause of magnetic reversals in oceanic basins. Usually the magnetic orientations recorded in oceanic basins correspond to those in terrestrial rocks of the same age.

2.6.2 *Depositional remnant magnetism*

This occurs in sedimentary rocks when magnetic minerals are orientated with respect to the Earth's magnetic field at the time of deposition. Once deposited, re-orientation is prevented by the surrounding and overlying sediments, unless they are heated to a temperature above the Curie point.

2.6.3 *Palaeo-magnetism and Polar wandering*

Palaeo-magnetism is a characteristic of the magnetic field preserved in rocks and minerals at the time of their formation.

By plotting the apparent positions of the magnetic poles for rock sequences of different ages on the same continent, it is possible to show either that the pole has wandered significantly or that the continent has moved relative to the magnetic pole. By comparing the apparent wandering curves constructed for different continents, it is possible to see how they have moved in relation to each other. It is also possible to determine how continents have joined together or split apart. If these data are combined with dating methods, it is possible to show how these changes have progressed through time. In other words, this information shows that Continental Drift has taken place (Chapter 8): in fact, it provided the first quantifiable data that could be used to prove that Continental Drift existed prior to the Mesozoic Era. This is because continuous magnetic anomalies in the oceanic crust only exist back to the Triassic Period, the start of the Mesozoic Era, which comprise the oldest known oceanic crust found in present oceanic basins.

An obvious question that arises from this is why is the oceanic crust relatively young compared to the continental crust? The answer is that all the oceanic crust that is older than this has been destroyed by subduction, which is one of the principal processes of Plate Tectonic theory (Chapter 8).

Background

For remnant magnetism to be valid, several assumptions are made:

1. The rock formation has not been rotated, folded, or re-magnetized since deposition.
2. The geomagnetic field has always been the same simple configuration as known today, i.e. that the magnetic poles have always been close to the Earth's rotational poles.

Difficulties may be encountered through:

1. *Secular variation.* This is the variation in the Earth's magnetic field over periods greater than five years. Magnetic readings have been recorded since the 16th century, because they were used as a primary method of navigation. These show that the magnetic pole has gradually moved – in London, for instance, it was 11°E of true North in 1580; 0° (true North) in 1660; 24°W in 1820; and 7°W in 1970. However, we only have detailed, reliable records that go back over about 300 years. These show that there has been a westward drift of about 0.2°/year in the position of the magnetic poles. Over a period of several thousands of years, it appears that the magnetic poles rotate around the rotation poles.
2. *Dipole field.* At present, this is approximately 11.5° from the rotational pole. One of the problems that this causes is that it is not possible to determine palaeo-longitudes (relative east-west positions), only palaeo-latitudes (relative north-south positions). This means that you

cannot detect any movement if an area of the Earth's surface has only moved in a longitudinal direction.

3. *True polar wander.* As the magnetic poles move over time, due to convection in the Earth's outer liquid core (true polar wander), any reconstructions of palaeo-magnetic data relative to the present position of the magnetic poles are referred to as "apparent polar wandering" (APW) paths, as they include components of movements of the Earth's surface and true polar wander. To be able to separate out these two, you need to be able to either:
   a. Determine the net motion of the entire Earth's surface relative to its axis of spin. This is known as the Vector-Sum method.
   b. *Mean-Lithosphere method.* This is based on the study of the random movement of individual plates compared to the average movement of the Earth's surface.
   c. *Hotspot method.* This compares the movement of plates to hotspots that are assumed to remain in the same position relative to the magnetic poles. One of the problems associated with this method is that many hotspots also appear to move, even though their movement is significantly slower in relation to the plates that move above them.
4. *Normal and reversed poles.* For a single palaeo-pole reading, it is difficult to tell whether it has a normal or reversed polarity. You need to have other data, such as the palaeo-climate or APW curves from other areas plus the age of the rocks, to be able to determine its polarity.
5. If the rocks have been significantly deformed after their formation, this can affect their magnetism. For this reason, the best data comes from flat-lying or gently folded sequences. If the rocks have been rotated, this information can be back-calculated. If the rocks have been strained (deformed), it is not always possible to determine the amount of deformation. If the rocks have been highly-deformed and partially metamorphosed, then their original magnetism can be overprinted by the magnetic field present at the time of deformation.
6. Finally, if a rock is composed of minerals that were formed at different times, i.e. different cooling rates, or through chemical precipitation over a long period of time, it may contain a variety of magnetic data.
7. We do not know why reversals occur.
8. We do not fully understand how they occur.
9. We are not sure how long they take to occur.

Discussion point

Do we really need to be able to date rocks precisely in order to have an idea of their age?

3
The origins of the geological time scale

3.1 Introduction

Before we look at the way in which the Earth's different rock units were identified, grouped together, and named, it might be useful to explain some of the terms that are used to help you understand how the observed geological units were subdivided.

As we saw in Chapter 2, rocks and the fossils they contain can be dated and therefore ordered in different ways. In its simplest form, they can be classified in four ways:

1. *Lithostratigraphy* (rock stratigraphy), which refers to the rock units;
2. *Chronostratigraphy* (time stratigraphy), which refers to the time periods in which they were formed;
3. *Geochronology*, which considers the "time units" in which rock sequences were formed;
4. *Biostratigraphy* (life stratigraphy), which is based on the fossil sequence.

It may seem odd to have four different ways in which rocks and fossils can be ordered, but unfortunately each one is based on recognizable points of change that do not necessarily correspond to each other.

To non-geologists who read this it may seem strange to be able to define sequences of rocks in different ways, but one set of names is based on rock types and sequences (lithostratigraphy) and the other is based on dating (chronostratigraphy) of the fossils, etc. that they contain, and, more recently, absolute dating.

Lithostratigraphic terms include Formation, which is a mappable rock unit with distinctive upper and lower boundaries. Formations can be divided into individual units that are called Members or Beds, or can be combined into larger Groups or Supergroups.

Time Matters: Geology's Legacy to Scientific Thought, 1st edition. By Michael Leddra. Published 2010 by Blackwell Publishing Ltd.

The fundamental time or chronstratigraphic unit is the System. This includes all the rocks that were deposited in a particular area (the stratotype) that can be extended to other areas. Systems can be subdivided into Series and Stages. The basic pure time unit is the Period, which can be subdivided into Epochs and Ages or combined into Eras, which have also been combined into longer periods of time called Eons (see the geological time scale (Table 1.1) at the end of Chapter 1). Also, chronostratigraphy and geochronology units can use the same names, i.e. the Carboniferous Period and the Carboniferous System.

Biostratigraphic units are defined by the presence of particular groups (assemblages) of fossils that are distinctive to a particular sequence of rocks, which can be placed into the chronostratigraphy and geochronologic sequence.

Discussion point

Before reading the following section, consider these questions:

Is it necessary to have a geological time scale?

As fieldwork was key to the establishment of the geological time scale, would you expect each unit to be investigated in a systematic or sequential order?

Would you expect to find that the sequence of rocks was repeatable and predictable, not only across the UK but also across Europe or even further afield?

If the answer to the last question is no, how would you establish links between different rock sequences?

The complex state of the stratigraphic units has meant that major divisions of the geological column such as the Devonian, Silurian, and Cambrian were not natural entities waiting to be discovered. It is important to realize that when the different divisions of the geological time scale were identified, the people who undertook this work were not building on, clarifying, or even refining an existing classification; they were effectively starting with a clean sheet. They looked at rock sequences not only across this country but also across Europe and Russia. Tying such dispersed locations together was no mean feat, and those that undertook this work should be held in the highest regard, especially as with over 180 years of further research their units have largely remained intact.

As you will see, the order in which the geological systems were established was by no means systematic. It should also be pointed out that although over 100 people were involved in sorting out the geological column, only about a dozen were of the "right social class, degree of commitment or expertise" or could get to London for the important meetings at the Geological Society and so were credited with the work.

One of the most remarkable facts in the study of the geological history of the Earth is that a significant part of it was established here in the UK. Why should the development of the geological time scale have been centred on Britain that, after all, is such a relatively small country?

Michael Bouter, in his book *Extinction, Evolution and the End of Man*, makes the following observation of the UK: "there are few parts of the world where it is possible to travel so far into geological time so quickly and see the sequence of rocks as they were laid down in the changing environments of the past".

The following is a very brief summary of the way in which the geological time scale and systems we use today were established. It is not intended to be comprehensive, but should be used as a guide for further investigation.

3.2 Jurassic

In 1795, Alexander von Humboldt (Chapter 2) looked at the rocks in the Jura Mountains (Calcaire de Jura) and called them the Jura–Kalkstein. He thought that they were just another formation of the Werner Neptunist scheme (see Chapter 4 for a discussion of Neptunism). As we have seen, between 1797 and 1815, William Smith produced geological successions and maps of England and Wales in which the detailed stratigraphy of the rocks of what were later to be defined as the Jurassic System were included. Some of these rocks were grouped together as the Oolitic Formation by Buckland in 1818, and the Oolitic Series that overlays the Lias (the original names used for the Lower Jurassic rocks in Britain) (Table 3.1) by Conybeare and Phillips in 1822, who grouped some of these rocks together as the Oolite Formation. These were found to be equivalent to the Jura–Kalkstein.

In 1829, Alexander Brongniart (1770–1847) – a French chemist, mineralogist, and zoologist – used the term, "*Terrains Jurassiques*" for Conybeare and Phillips' Lower Oolitic Series (in the UK the name, "Jurassic" co-existed with that of the Lias and Oolite for many years). One reason why

Table 3.1 The Jurassic Period (values are × 1 million years)

Eon	Era	Period	Epoch	Date
				144
Phanerozoic	Mesozoic	Jurassic	Late	159
			Middle	180
			Early	206

Smith and others were able to identify and subdivide the British Jurassic sequence was that the ammonites they contained evolved so rapidly and were so abundant in Jurassic rocks, that they could be used to subdivide the Jurassic System into very small units compared to most other geological sequences.

In 1839, Christian Leopold Von Buch (1774–1853), a German geologist and palaeontologist who studied with Von Humboldt and Werner, redefined the Jurassic as a system in its own right and dedicated the paper in which he outlined his ideas to Werner.

3.3 Carboniferous

The name "Coal Measures" was proposed as a name by geologist John Farey (1766–1826) in 1807 and was included in a three volume report on the agriculture and minerals of Derbyshire, published between 1811 and 1817. This covered the sequence of rocks in which the British coal seams were contained. English geologist John Whitehurst (1713–1788) defined the Millstone Grit (another major division of the Carboniferous, found below the Coal Measures) in 1778. In 1822, William Daniel Conybeare (1787–1857) (Fig. 3.1) and William Phillips (1775–1828) grouped all the

Fig. 3.1 **William Daniel Conybeare**

related strata together and proposed that the Coal Measures, Millstone Grit, and the Mountain or Carboniferous Limestone found in the North of England, which lies above the Coal Measures, should be classified as Carboniferous in age (Table 3.2). In 1835, John Phillips (1800–1874) proposed the name "Carboniferous System". This was one of the first systems to be proposed based on fossil correlations rather than just rock types. It is interesting to note that John Phillips' father married William Smith's sister and it was he who championed Smith's case to right the wrongs done to him by the establishment (Chapter 2).

3.4 Triassic

In 1834, whilst studying the local rocks, Fredrich August von Alberti (1795–1878) – a German palaeontologist who lived in the salt-mining area of Germany – found that the local geology comprised a very clear three-fold division of rock types named Bunter Sandstones, the Muschelkalk Limestone, and the Keuper Marls from the Calcareous Alps in Austria. He proposed the name Triassic System for this sequence (Table 3.3). One of the problems with the type section (the sequence of rocks which define the Triassic, located in Southern Germany) was that it contained few fossils, even in the limestone, that were marine in origin. It was also discovered that this three-fold sequence was limited in extent. Due to extensive folding and faulting, the Ammonite Zones, on which the dating of the sequence was based, were later found to be incomplete and not always in chronological order.

Table 3.2 The Carboniferous Period (values are × 1 million years)

Eon	Era	Period	Epoch	Date
				290
Phanerozoic	Palaeozoic	Carboniferous	Late	323
			Early	354

Table 3.3 The Triassic Period (values are × 1 million years)

Eon	Era	Period	Epoch	Date
				206
Phanerozoic	Mesozoic	Triassic	Late	227
			Middle	242
			Early	248

In the UK, the Triassic, known for the New Red Sandstones, comprised rocks formed in desert conditions when the country was located nearer to the equator. Consequently, fossils are rare, which made establishing the age of the sequence difficult.

3.5 Tertiary

This is the only one of Arduino's terms (Chapter 2) still in general use. Originally, Arduino defined it in Italy in 1759 but the type sections were established in the Paris Basin. The Eocene (*eo* means dawn or new life, and *cene* means recent), Miocene (*mio* meaning less recent), and Pliocene (*plio* meaning more recent), which are all subdivisions of the Tertiary (Table 3.4), were defined by Charles Lyell in 1833, based on the proportion of living to extinct fossil types that they contained. Their proportions decrease with increasing time, thus in the Eocene rocks, this was 3%; in the Miocene, it was 17% and in the Pliocene, it was between 50 and 67%.

At the time, Lyell (Chapter 5) and his contemporaries used the name "Neozoic", meaning more recent than the Palaeozoic – to cover all the rocks younger than the Permian in age. John Phillips later divided these rock sequences into the Mesozoic and Cainozoic (now Cenozoic) in 1841. Later still, studies established extensive marine, brackish, freshwater, and continental sediments in Northern Europe between Lyell's Eocene and Miocene, for which Heinrich Ernst von Beyrich (1815–1896) – a German geologist – proposed the name "Oligocene" in 1854. The name Oligocene is based on the Greek words for "few" and "new", which refers to the reduction in the number of new mammals that evolved during this epoch compared to those found in Eocene strata.

In 1874, Wilhelm Philipp Schimper (1808–1880), who was Professor of Geology and Natural History at the University of Strasbourg, defined the

Table 3.4 The Tertiary Period (values are × 1 million years)

Eon	Era	Period		Epoch	Date
					0.01
Phanerozoic	Cenozoic	Tertiary	Neogene	Pleistocene	1.8
				Pliocene	5.3
				Miocene	23.7
			Palaeogene	Oligocene	33.7
				Eocene	54.8
				Palaeocene	65

early Eocene rocks in Western Europe as Palaeocene (which means "old to new" or "early dawn of the recent"), based on his studies in the Paris Basin. A two-fold division of the Tertiary (which is regarded as a sub-era), into the Palaeocene (*palaeo* meaning ancient) and Neogene (*neo* meaning newborn), was based on the climax of the Alpine Orogeny (an "Orogeny" is a period of mountain building) and was introduced in 1853 by Moritz Hornes (1815–1868), an Austrian palaeontologist. The earliest period (Palaeogene) contains the Palaeocene, Eocene, and Oligocene and the later period, the Neogene, contains the Miocene and Pliocene (Table 3.4).

3.6 Cambrian

Adam Sedgwick (1785–1873) was born in Dent, Yorkshire in 1785. In 1817, he was ordained into the Church of England and in 1818 became Woodwardian Professor of Geology at Cambridge, where he has been described as "a plain-spoken, respected gentlemanly geologist". In 1822, he started to map the geology of the Lake District and from 1829 to 1830, he became President of the Geological Society.

Using the mapping techniques established by William Smith, Sedgwick started to map the rocks of north and central Wales in 1831 (Fig. 3.2).

Fig. 3.2 **Adam Sedgwick**

Table 3.5 The Cambrian Period (values are × 1 million years)

Eon	Era	Period	Epoch	Date
				490
Phanerozoic	Palaeozoic	Cambrian	Late	501
			Middle	513
			Early	543

Originally Murchison (see below) and Sedgwick had intended to map the rocks in Wales together as these represented rocks that held the earliest known forms of life that lay below, and were therefore older than the Old Red Sandstone of the Devonian System (see below). However, this joint venture fell through and both geologists went their separate ways.

One of the problems that Sedgwick encountered in Wales was that the rocks towards the base of the sequence contained few fossils. He therefore had to base his mapping primarily on the characteristics of the rocks. In fact, it has been said that "it is to Sedgwick's credit that he recognized and named the Cambrian (Table 3.5) almost a century before the best fossils were found". As Palmer says, "we now know that Sedgwick had taken on a task that could not really have been satisfactorily concluded given the state of knowledge at the time".

During one of his summer expeditions, Sedgwick took a young graduate from the University of Cambridge named Charles Darwin with him.

3.7 Silurian

Roderick Impey Murchison (1792–1871) (Fig. 3.3) was born in Tarradale, Scotland. After attending Durham School, he joined the army and eventually served under Sir John Moore, who is famous for establishing the Light Infantry regiments of the British army. After eight years he left the army, married, and went to live in Barnard Castle, County Durham. It is said that it was his wife and Sir Humphry Davy (inventor of the mining safety lamp) who persuaded him to take up geology instead of foxhunting and shooting. He appears to have thrown himself into this new interest with great enthusiasm and went on to study the rocks in Sussex, Hampshire, and Surrey before joining Lyell to investigate volcanic rocks in France and other sequences in Italy, Germany, and Switzerland.

In 1831, apparently on the advice of William Buckland, he started to look at the rocks around the borders of England and Wales to see whether the vast sequence of Welsh greywackes (mudstones and muddy sand-

Fig. 3.3 **Roderick Impey Murchison**

stones) and the overlying red sandstones (later they would be known as the Old Red Sandstone) were related. He started to study these "transition rocks", using their fossils as a method of correlation. He worked from mid-Wales northwards. At the same time, Adam Sedgwick, a close friend, was working on the lower part of the sequence. In 1839, Murchison subsequently proposed that the upper sequence of Welsh greywackes should be grouped into the Silurian System (the name is derived from the Silures, a Welsh tribe who had lived in the area).

After presenting their systems in a joint paper in 1835, Murchison tried to claim ownership of Sedgwick's Cambrian sequence based on the fossil evidence he found in his own rocks.

It is generally accepted that Murchison effectively tried to "land-grab", and take over Sedgwick's Cambrian rocks as part of his own Silurian System. He was accused effectively of a campaign of intellectual imperialism, as he wanted as much territory in his Silurian System as possible. It is reported that when presenting a paper on the Silurian, someone else commented, "I can foresee the fate of geology for the next eight years – half of the globe will become Silurian". He is also said to have gained a "well-earned reputation for browbeating his colleagues". In 1893, Albert Auguste Cochon de Lapparent (1839–1908), a Professor of Geology and Mineralogy at the Catholic Institute in Paris, proposed the name "Gotlandian" for the

Table 3.6 The Silurian Period (values are × 1 million years)

Eon	Era	Period	Epoch	Date
				417
Phanerozoic	Palaeozoic	Silurian	Late	423
			Early	443

post-Ordovician Period (Murchison's sequence) to compete with the name "Silurian", but this was finally dropped in 1963. The Silurian Period (Table 3.6) was the first to have all its major divisions agreed internationally with standardized reference points and boundary stratotypes.

Due to his position, Murchison held much sway and power, and as Palmer describes:

> He had a very personal proprietorial view of "his" system, typically referring to the landscapes of "my Silurian Region" or to fossils as "my published Silurian types". He developed the argument that "the great mass of rocks which Sedgwick had called Cambrian, but without defining their fossiliferous contents, were nothing but replications, in a more altered and slaty condition, of my Silurian types". Murchison was wrong, Sedgwick's Cambrian strata do contain separate and distinct fossils, but Sedgwick was partly to blame because of his long delay in describing them.

At this time, maps played an important role in being able to record and define changes in rock type and fossils. Once a period had been defined, large sections of the Earth could be coloured in as belonging to that period. Murchison – who was a very enthusiastic President of the Geological Society between 1831–1832 and also in 1842–1843 – wanted to have as much of the globe "conquered" by British science as possible, and this meant making sure that the world's rocks were defined by British terms. He was convinced that Britain's greatness as an industrial power had been predestined, because her rocks contained abundant supplies of coal and iron. Palmer portrays a different side to his character in noting that "Murchison was always keen to try to apply geological knowledge to economic ends", for instance, he "made considerable efforts to deter landowners, ignorant of the geological constraints, from wasting money searching for coal in rocks that would never yield any".

Interestingly, as reported in my book *Turn and Burn: The Development of Coal Mining and the Railways in North East England*, William Smith had almost the reverse problem. From his understanding of the nature of stratigraphy he knew that the coal seams exposed to the west of the limestones in County Durham must continue eastwards. Local knowledge at the time held the view that the coal stopped at the edge of the limestone escarpment. He tried, and eventually succeeded, in persuading land- and

Fig. 3.4 **Claxheugh Rock, Tyne and Wear. This cliff-face, exposed along the bank of the River Wear, shows the problems the mineowners faced. The upper section comprises a reef in the Permian Magnesian Limestone (L), which overlies the Permian Yellow Sands (S). The Carboniferous Coal Measures (CM) are exposed at river level in the lower left-hand corner of the photograph**

mineowners in County Durham to dig a mineshaft through the Permian limestone south of Sunderland to find coal (Fig. 3.4).

3.8 Devonian

In 1839, Murchison and Sedgwick jointly named the Devonian Period from rocks they had studied together in Devon (Table 3.7). This actually proved to be a poor type area due to intense deformation caused by the culmination of the Hercynian Orogeny that ended during the later part of the Carboniferous Period. (A type area is an area that contains rocks and fossils typical of the sequence.) One of the problems they encountered was that the base of the sequence was not exposed. However, the rocks and the fossils they contained were distinctive and were later used in the Rhineland in Germany to determine that the rock sequence there – which was better exposed – was also Devonian in age.

The Devonian Period was said to have been created as a means of resolving a dispute, generally known as the "great Devonian controversy",

Table 3.7 The Devonian Period (values are × 1 million years)

Eon	Era	Period	Epoch	Date
				354
Phanerozoic	Palaeozoic	Devonian	Late	370
			Middle	391
			Early	417

Fig. 3.5 **Henry De La Beche**

between Murchison and Henry De La Beche (1796–1855) (Fig. 3.5). This dispute, which lasted from 1834 to 1842, is said to have started as a minor problem over the dating of the Devonian strata, and ended with a new view of the history of the Earth.

This represents a fundamentally important period of the Earth's history, as it marks the time when life "flourished in the sea and plants and vertebrates became diverse and abundant on the land". It represents the time when fish, plants, and trees evolved; in other words, it was a period when life on Earth changed significantly. For the first time, life moved out of the seas and onto the land. Therefore, establishing its position in the geological sequence was vital. The lack of fossils in most of the British Devonian, which was generally termed the Old Red Sandstone, meant that establishing whether the rocks in Devon (that were fossiliferous) were of the same age was crucial.

Background

Rocks of the Palaeozoic Era (i.e. the Cambrian, Ordovician, Silurian, Devonian, Carboniferous, and Permian periods) posed problems for a number of reasons:

1. They were usually intensely folded and faulted, which made sorting out the stratigraphy a problem.
2. They were frequently metamorphosed, and much of their fossil record had been destroyed.
3. Much of the sequence consisted of greywackes or desert sandstones.
4. Large sections of the "layer cake" Palaeozoic rocks were missing, which meant that tying the sequence together became problematic.
5. Fossil evidence appeared to show that some of the rocks were "out of sequence" compared to the expected order.

The point of the argument was that the rocks appeared to be both Silurian (the period before the Devonian) and Carboniferous (the period that follows the Devonian) in age. The controversy began in 1834 when Henry De La Beche, one of the founding members of the Geological Survey in 1835, found plant fossils in greywackes in Devon. Murchison, who had not seen the rocks themselves, immediately said that De La Beche must be wrong because "greywacke were old rocks" and the plant fossils were "young". (Note that Murchison used "old" and "young" as relative terms.) At the time, everyone had thought that the rocks in Devon, being marine in origin, were approximately the same age as the ones in Wales, i.e. Silurian or Cambrian. However, De La Beche had found coal deposits in them, which are invariably associated with much younger Carboniferous rocks. This immediately caused a problem, as no plant life had been found in any of the older Welsh rocks. The presence of coal in Devon implied vegetation growing on the land surface, something that had not been seen in such apparently old rocks.

It was realized that the Carboniferous rocks and fossils were at the top of the "layer cake" and that rocks such as greywackes found in the Cambrian and Silurian were near the bottom. Murchison therefore claimed that the rocks that De La Beche had found had to be close to the top of the sequence because of the plant fossils they contained, and that therefore they could not be somewhere in the middle as De La Beche had proposed. He thought that the sequence must contain an unconformity (a break in deposition of the rocks) if relatively young fossils, which looked like those of the Carboniferous Period, lay on top of, and close to, older "Silurian" fossils. (Until about this time, rock type had generally been used for dating rocks, hence Murchison thought that the greywackes were "old rocks".) By claiming that Coal Measure plant fossils could be found in apparently Silurian or older rocks, Murchison believed that De La Beche was taking a step backwards, because this implied a lack of faunal succession.

De La Beche admitted that he had mismapped the Devonian structures and that the rocks did exist close to the top of the "rock pile". He still claimed that they were closer to the Silurian than the Carboniferous in age, as he had found no unconformities in the sequence, even though his examination of the plants in the coal showed them to be almost identical to those found in younger Carboniferous coals. In 1836, Murchison and Sedgwick mapped the sequence in Devon, looking for any unconformities that might solve the problem. Eventually, Murchison agreed that no unconformities existed but insisted, based on fossil evidence, that the rocks could not belong to his Silurian Period either. A further problem the British geologists encountered was that the rocks in Devon, and their equivalents in Europe, looked nothing like the Old Red Sandstone found under the Carboniferous rocks elsewhere in Britain, which usually comprise desert sandstones and conglomerates (pebble beds).

Finally, it was realized that the rocks were neither Silurian nor Carboniferous in age, but represented a previously unrecognized period of time (Table 3.7). They were later identified as being equivalent in age to the Old Red Sandstone (a well-known rock unit that contained freshwater fish fossils), which of course were difficult to correlate with the marine fossils found in the sequence in Devon. In 1839, Murchison decided that the top of the sequence was conformable (parallel and in sequence) with the overlying Carboniferous. He also had to acknowledge that different fossil assemblages could occur contemporaneously in different types of rocks; in other words, fossils were not necessarily confined to one particular rock type.

Yet again, it appears that status and politics may have played at least some part in this dispute. De La Beche was trying to protect his status as Director of the Geological Survey (the controversy threatened to destroy his credibility as a geologist) and Murchison, who succeeded him as Director, was trying to use his position to gain an advantage over Sedgwick. Murchison also had an advantage over De La Beche, because he was wealthy and could afford to travel around the country to attend important meetings to make his case.

Palmer points out that De La Beche had not always had to work for a living. Originally, he had inherited a sugar plantation in Jamaica before returning to England in 1824. However, following the abolition of slavery, his income from the plantation shrank to such an extent that he had to get a paid job, mapping the geology of Devon – where he lived – for the Government. This meant that when he was on fieldwork, De La Beche could not attend the meetings to present his work.

Eventually, a sub-division of the system was established, based on marine fauna such as conodonts (microfossils that look similar to teeth or small combs) primarily found in the Ardenne-Rhenish area. This meant that, for a time, the sequence was known as the Rhenian, a name that was preferred by some workers to that of "Devonian", as it reflected an area where good fossil correlation could be established.

It can be seen that it was, to some extent, the power struggle within the geological establishment at the time that led to the important "discovery" of the Devonian succession.

3.9 Permian

Palmer records that it was Murchison's social skills, networking, and contacts that enabled him to plan his expeditions to Russia. Following these, in 1841, he named the Permian System after the sequence of rocks he and his Russian co-workers had investigated in the Perm area of Russia. In fact, Palmer points out that "he could not have asked for a better remit and Murchison's passport was endorsed by Czar Nicholas I himself". They established that these rocks were above the rocks of the Cambrian System and they thought that they were the same general age as those of the English Magnesian Limestone and New Red Sandstone (these were originally identified in 1822 but not dated at that point in time), as well as the Rotliegendes and Zechstein sequences found in Germany (which were mapped around 1808). Murchison thought that even if the German rocks did not match those found in Britain, the fossil succession should be broadly the same (Table 3.8). The problem with comparing the British and German rocks was that, because they were limestones, dolomites, and sandstones, including desert sandstones, they contained few, if any fossils that could be correlated over large areas. Therefore, as the rocks could not be correlated with those in other areas, it was difficult to tie them into the Russian and other sequences. This also meant that it would be difficult to justify the creation of an entirely new geological system in Western Europe, which may or may not correspond to those found further afield.

Discussion point

It is interesting to note that Murchison is said to have conducted his tour of Russia "at a gallop and geologized while his carriage crossed the Urals behind horses sweating under the leash". These rocks contained a "vast series of beds of marls, schists, limestones, sandstones and conglomerates". Most geologists today would envisage the geological "greats" as studying the rocks carefully and in great detail. Murchison's speed was partially to ensure that his Permian was established and into print before George Reman, a German physicist and explorer, could establish his own name for the same rocks.

Table 3.8 The Permian Period (values are × 1 million years)

Eon	Era	Period	Epoch	Date
				242
Phanerozoic	Palaeozoic	Permian	Late	256
			Middle	290

3.10 Mississippian

In 1870, Alexander Winchell (1824–1891), a Professor of Geology and Palaeontology at the University of Michigan, was the first person to name "Mississippian" for the Lower Carboniferous rocks in the Mississippi Valley near St Louis. These are equivalent to the Mountain or Carboniferous limestones of Britain and Europe (Table 3.9). After further mapping of what turned out to be a complicated sequence, Henry William Shaler (1847–1918), an American palaeontologist and geologist with the US Geological Survey (USGS), proposed the name Mississippian Series in 1891 for the rocks. Although the sequence is predominantly limestone, the American sequence is significantly different to that in Europe, as the Mississippian and the overlying Pennsylvanian (see below) are separated by a significant unconformity. It was the identification of this unconformity that led Chamberlin and Salisbury to raise both the Series to the level of Systems in 1906.

Although there is an international agreement that the two American divisions, Mississippian and Pennsylvanian, should be recognized and may be used across the world, the boundary between the two falls within the lower half of the Upper Carboniferous rather than at the boundary between the Upper and Lower Carboniferous in our sequence of rocks. It is therefore unlikely that the American systems will ever be used in Britain and Europe.

Table 3.9 The Mississippian Epoch (values are × 1 million years)

Eon	Era	Period	Epoch	Date
				290
Phanerozoic	Palaeozoic	Carboniferous	Pennsylvanian	323
			Mississippian	354

Table 3.10 The Quaternary Period (values are × 1 million years)

Eon	Era	Period	Epoch	Date
				0.0
Phanerozoic	Cenozoic	Quaternary	Holocene	0.01

3.11 Quaternary

This is the last, most recent period in the Earth's history, which has been given various names in the past. In the early years of the 19th century, the term "Alluvium" was used. In 1823, it was thought that the older term Diluvium, which had been used for rocks and sediments produced by the Biblical Flood (Chapter 5), was an appropriate term for the Quaternary sequence.

In 1829, a French geologist, Jules Pierre François Stanislaus Desnoyers (1800–1887), proposed the name Quaternary (meaning fourth) for rocks in the Seine Basin that were thought to be "very young".

In 1837, the term Drift was being used for sands, gravels, and boulder clay that were thought to have been deposited by floating ice. In 1839, Charles Lyell proposed the Pleistocene Period (meaning "most new" in Greek) for the ice-age deposits, which post-dated the Pliocene sequence that had already been established. Around 1840, it was recognized that the widespread glacial erratics found in the Alps and northern Europe were the result of former ice movements (Chapter 5). In 1854, Karl von Adolf Morlot (1820–1867), a Swiss geologist and archaeologist, proposed that the Quaternary Period (Table 3.10) should include both the Pleistocene (*plei* meaning most) and Lyell's "recent" deposits of postglacial age (which were later named "Holocene" in 1885).

3.12 Ordovician

The rocks below the Old Red Sandstone, close to the top of the "layer cake" and the older crystalline rocks that formed the base, were regarded as a transitional sequence. In 1835, Sedgwick and Murchison had subdivided this into separate divisions that they then tried to include in their own Cambrian and Silurian systems.

Initially Sedgwick and Murchison discussed their results for the establishment of the Cambrian and Silurian systems without any disagreement. Later, when it became obvious that there was an overlap in the fossils they contained, a major conflict developed. It is said that if Murchison had studied the fossils in his sequence more carefully, he might have relinquished some of his Silurian strata to Sedgwick's Cambrian and therefore they would probably have avoided the ensuing argument and remained friends. However, Sedgwick attacked Murchison in print and Murchison was incensed: their long-term friendship ended and it is said that they did not speak to each other ever again. This dispute between two such prominent geologists was considered so bad that resolving the problem of the apparent overlapping sequences was not settled until after their deaths.

These problems arose because Murchison, who was working from the top of the sequence downwards, was looking at his Silurian rocks, which contained "plenty" of fossils, whereas Sedgwick's Cambrian rocks contained fewer fossils that he could use for correlation. This resulted in Murchison claiming the same rocks in his Lower Silurian that Sedgwick was including in his Upper Cambrian; in fact, Murchison also claimed that the greywackes in North Devon (Devonian) were also part of his sequence. The more that people looked at Sedgwick's Cambrian sequence, the more inconsistencies they found. This meant that bit by bit Sedgwick was losing his Cambrian sequence, and Murchison eventually claimed that it should all be part of his Silurian.

In 1853, Sedgwick found a layer of rocks that looked similar to his Cambrian rocks, but which were above and therefore younger than some of the rocks that Murchison claimed were Silurian in age.

The basic problem they both faced was that the upper and lower boundaries of the disputed sequence were marked by unconformities. The rocks were also highly deformed because they had been extensively folded in the later Caledonian mountain-building event. In some places this deformation had been so severe that some of the sequence had been overturned, a situation which subsequent geologists were able to identify through a careful study of the fossil sequence.

In 1879, Charles Lapworth (1842–1920) (Fig. 3.6), who trained as a teacher in the Southern Uplands of Scotland before becoming a geologist, proposed the name Ordovician for the disputed rocks that comprised 3,600 m of volcanic and sedimentary rocks (Table 3.11). He set the boundaries based on variations in the graptolite fossils they contained. He also realized that the fossils in the disputed rocks were actually different to those in both the Cambrian below and Silurian above. Some people say that he chose the name Ordovician because the rocks occur in the area in which the Ordovices, a Welsh tribe, had lived. Others say he may have chosen it because it was a war-like tribe and he was therefore marking the battle between the two great geologists.

Fig. 3.6 **Charles Lapworth**

Table 3.11 The Ordovician Period (values are × 1 million years)

Eon	Era	Period	Epoch	Date
				443
Phanerozoic	Palaeozoic	Ordovician	Late	458
			Middle	470
			Early	490

The establishment of the Ordovician System meant that Murchison's Silurian System shrank considerably and that Sedgwick's Cambrian System was re-established. In fact, as Palmer points out:

> Throughout the 20th century, the establishment and international acceptance of the Ordovician and the re-instatement of the Cambrian also led to a reappraisal and posthumous increase in Sedgwick's status as an important if somewhat tragic figure in the history of geology.

Palmer also adds that although Murchison was, and still is, largely portrayed as a snobbish and pompous empire builder who unfairly did Sedgwick "down":

> ... his personality and driving force might have been overbearing at times, but it is also possible that to begin with he inspired Sedgwick to achievements such as their work in Devon and in the Alps that Sedgwick might not have achieved on his own.

3.13 Cretaceous

Belgian geologist Jean Baptiste Julien d'Omalius d'Halloy (1783–1875) (Fig. 3.7) proposed the name Cretaceous in 1822 for the rocks around the Paris Basin that contained formations of "chalk with its tufas, its sands, and its clays". It was originally based on the Chalk, but was then extended to cover the other "terrains" recognized by d'Halloy. William Smith had already mapped four strata in England between the "Lower Clay" (Eocene) and the "Portland Limestone" (which is Jurassic in age), namely the "White Chalk, Brown or Grey Chalk, Greensand, and Micaceous Clay or Brick Earth". In 1822, Conybeare and William Phillips proposed that the chalk formed the Upper Cretaceous, with the rest forming the Lower Cretaceous (Table 3.12).

Fig. 3.7 **Jean Baptiste Julien d'Omalius d'Halloy**

Table 3.12 The Cretaceous Period (values are × 1 million years)

Eon	Era	Period	Epoch	Date
				65
Phanerozoic	Mesozoic	Cretaceous	Late	99
			Early	144

3.14 Pennsylvanian

In Britain, this is equivalent to the Millstone Grit and the Coal Measures. In Pennsylvania, the Pennsylvanian Group (which includes the Coal Measures) is primarily marine in origin and is more fossiliferous than the European equivalent, which is deltaic. It has therefore been divided into Lower, Middle, and Upper (as is the Mississippian) (Table 3.13).

Table 3.13 The Pennsylvanian Epoch (values are × 1 million years)

Eon	Era	Period	Epoch	Date
				290
Phanerozoic	Palaeozoic	Carboniferous	Pennsylvanian	323
			Mississippian	354

3.15 Proterozoic

When Sedgwick proposed the Palaeozoic Series in 1838, he named the underlying rocks the Primary Stratified Groups. He said, when they were found to contain organic remains, that they should be termed the Protozoic System, which means "early life". In 1841, when Phillips extended the meaning of Palaeozoic to include the Cambrian to Permian sequence, he introduced the terms "Hypozoic" or "Prozoic" for the pre-Palaeozoic Rocks, i.e. those of the Precambrian. "Proterozoic" was first used by Emmons in 1888 and became established by Charles Richard Van Hise (1857–1918) in 1892.

Joseph Beete Jukes (1811–1869), a geologist who studied under Sedgwick, established the Precambrian in 1862, and in 1877 Henry Hick (1837–1899), a Welsh doctor and geologist, divided the British Precambrian into four units: the Lewisian in Scotland and the Dimetian, Arvonian, and Pebidian in Wales.

Table 3.14 Divisions of the Precambrian (values are × 1 million years)

Eon	Period		Date
			543
Precambrian	Proterozoic	Late	900
		Middle	1,600
		Early	2,600
	Archean	Late	3,000
		Middle	3,400
		Early	3,800
	Hadean		4,600

Table 3.15 Establishment of the geological time scale (based on Palmer, 2005)

Period	Founder	Era	Founder
Holocene	Gervais, 1867	Quaternary	Desnoyers, 1839
Pleistocene	Lyell, 1839		
Pliocene	Lyell, 1833	Cainozoic	Phillips, 1840
Miocene	Lyell, 1833		
Oligocene	Beyrich, 1854		
Eocene	Lyell, 1833		
Palaeocene	Schimper, 1874		
Cretaceous	D'Halloy, 1822	Mesozoic	Phillips, 1840
Jurassic	Von Buch, 1839		
Triassic	Von Alberti, 1841		
Permian	Murchison, 1841	Palaeozoic	Sedgwick, 1838 and Phillips, 1841
Carboniferous	D'Halloy, 1808 and Conybeare and Phillips, 1822		
Pennsylvanian	Williams, 1891		
Mississippian	Winchell, 1870, Williams, 1891		
Devonian	Murchison and Sedgwick, 1839		
Silurian	Murchison, 1835		
Ordovician	Lapworth, 1879		
Cambrian	Sedgwick, 1835		
Precambrian	Jukes, 1835		

3.16 Archean and Hadean

As the study of the geological sequences and the fossils they contain continued to develop, particularly during the last 50 years, it became apparent that life existed further back into the Precambrian than was previously thought. This, together with the discovery of increasingly older rocks, meant that the Precambrian as a unit was too broad. It represented nearly 90% of the geological time scale. It has therefore been subdivided into two Eons, the Archean and the Hadean (Table 3.14). The name "Archean" is derived from the Greek for "ancient"; "Hadean" is derived from the Greek for "unseen" or "hell". The Archean covers the earliest rocks so far found, whilst the Hadean covers the period of Earth history when the planet was originally forming.

Discussion point

If you consider the way in which the sequence of rocks and the geological time scale were established, it seems surprising that it has remained virtually unaltered since it was set out over a 50-year period in the middle of the 19th century (Table 3.15). However, in the majority of cases the major subdivisions, as listed above, were defined by careful examination and comparison of different sequences primarily throughout the UK and Europe. Subsequent investigations have mostly identified more complete sequences elsewhere, and consequently the majority of "type sections" are no longer found in the areas where the units were originally defined; yet the names have remained. This would have pleased their founders – particularly Murchison and Sedgwick.

If you were to look at geological maps of the UK, which show the exposures of the rock sequences for each of the geological periods in the order in which they were defined, it would be difficult to determine the overall picture of the progression of geological history at the time. This makes the achievements of those earliest geologists who, for whatever reason they undertook the work, even greater.

It is interesting to see how established geological ideas are still being distorted, as "creation scientists" try to re-interpret the geological time scale in order to put it into a flood-time perspective. In a recent article, the authors claimed that the geological time scale had been devised by people using the concept of evolution and implied that this was the reason that the rock sequences and therefore the geological periods were placed in the order we now use.

This chapter shows that nothing could be further from the truth. Most of the time scale was established **before** Darwin's book was published

(Chapters 5, 6, and 7), and the majority of the people involved in its establishment did not hold evolutionary views – in fact, a number argued against them. It is important to remember that most of those responsible for establishing the time scale and its divisions were **not** uniformitarianists – the idea hardly existed at the time. This means that the foundation of the geological time scale was **not** based on the premises of slow, steady-state processes operating over long periods of time, and with the introduction of uniformitarianism (Chapter 5), it did not change. Therefore, the original work was even more robust than creationists, and sometimes fellow geologists, acknowledge.

Having looked at the way in which geological history was divided up, consider the following questions:

What role do you think Sedgwick and Murchison really played in the development of the geological time scale and to what extent did "intellectual imperialism" affect the way in which it was constructed?

Do you think that their actions helped or hindered the development of the geological time scale?

Is the following statement really true: "the geological time scale is the most important contribution geology has made to the development of science"?

Further reading

If you would like to find out more about the establishment of the geological time scale, I would recommend reading the first part of Douglas Palmer's book, *Earth Time: Exploring the Deep Past from Victorian England to the Grand Canyon.*

In this text, Palmer notes that "there is a discrepancy between our knowledge of the history of practical geology in Britain and the development of its more academic theories". He points out that the Industrial Revolution, based on industrial mining and use of geological resources, began during the latter half of the 18th century and asks, "how is it possible that it should have succeeded so spectacularly in Britain when it was apparently in advance of the theorizing and published maps and information about the distribution of geological rock materials of economic value"? It is obvious that many people were involved in finding and extracting the geological resources, and between them, they held a body of knowledge that was "outside of the gentlemen cliques of the metropolis and the few universities that 'indulged' in science".

4
Plutonism versus Neptunism

4.1 Introduction

This chapter explores the battle between two of the most important opposing views in geology in the 18th century. This battle and its consequences have influenced, and continue to influence, the study of geology. It has had a profound effect on our view of the world – not just on how it was formed and how it has changed over time, but the way in which we view natural processes and speed of change. Various groups are still using many of the arguments today, even though our knowledge has moved on. Only since the early 1990s have some of the effects of this "battle" begun to change.

Discussion point

The conventional image of late 18th century geology is of a scientific battleground between the Plutonists who are regarded as the forward-looking, progressive thinkers, and the conservatives – the Neptunists. The Plutonists are generally regarded as the group that used scientific reasoning and principles to understand the formation of the Earth, whereas the Neptunists are usually portrayed as relating everything to biblical accounts and were therefore holding science back. This is a view that still persists.

Some of the "standard" views regarding the Plutonists and the Neptunists are outlined below. These clearly support those given above. However, these are then followed by a more complete review of the two theories, and their main followers, in which it is clear that many of the "main players" on each side held both "advanced" and "retrograde" ideas.

One of the most important people in the Plutonists' camp was James Hutton, whom we encountered in Chapter 1. He thought that all changes

Time Matters: Geology's Legacy to Scientific Thought, 1st edition. By Michael Leddra.
Published 2010 by Blackwell Publishing Ltd.

Fig. 4.1 **Arthur's Seat, Edinburgh**

affecting the Earth's surface had taken place slowly and gradually and published the conceptual framework for what would become uniformitarianism (Chapter 5). He based his ideas on the speed of erosion and deposition that he observed in his native Scotland. He thought that given enough time, present rates of activity would be sufficient to produce all of the features found in rocks. He also thought that the observable relationships and configurations in the landscape that he saw around him were the same as those that operated in the past. Hutton wrote:

> From the top of the mountain to the shore everything is in a state of change … We have a chain of facts which clearly demonstrate that the materials of the wasted mountains have travelled through rivers … There is not one step in all this process that is not to be actually perceived … What more can we require? Nothing but time.

He also studied evidence of volcanic activity in places such as Arthur's Seat in Edinburgh (Fig. 4.1) and developed the idea that the rocks he observed there were formed from molten materials originating from below the Earth's surface (Fig. 4.2).

4.2 Neptunism

The opposing view, Neptunism, was headed by Abraham Gottlob Werner and was based on the retreating ocean concept that, in later years, was considered to be an "old" or "regressive" idea. As we saw in Chapter 2, Werner believed, because of his mineralogical and mining background, that all rocks were precipitated out from a primeval, retreating ocean. Unfortunately, due to his influence and reputation, his ideas were then adopted, particularly in the UK, as a defence of the Biblical Flood catastrophism concept that he neither believed nor upheld.

Fig. 4.2 **The volcano at the Heads of Ayr (left), and volcanic bombs and debris at Dunbar (right top and bottom), Scotland**

Discussion point

It has been stated that the view of a battleground has been generated to conform to the idea of scientific progress, i.e. that scientific ideas progressively develop and, using hindsight, it is possible to find the links in the chain.

The Neptunists were not backward in their thinking and Werner did not believe in the Genesis Flood. Equally, some of the Plutonists held views that could hardly be regarded as progressive.

We have already seen that Werner and his contemporaries undertook a great deal of valuable work that helped establish some of the principles that are still used in present-day geological thinking. It was only their interpretation of the processes for the formation of rocks based on available data and established ideas, originally collected in a relatively small area, that were wrong.

In the latter half of the 18th and the first half of the 19th century, Neptunism became the standard model that virtually everyone used to explain the formation of rocks and an apparent decrease in sea level. It was thought that all rocks had crystallized out of some form of primordial "soup", with the Primary rocks – granites, basalts, and other dense rocks, which formed the highest mountains, forming first. Limestones, sandstones, and other Secondary rocks formed on middle and lower slopes as they crystallized out of the mineralogically depleted soup. Finally, muds and alluvium – the Tertiary formations – found in the bottom of valleys and low-lying planes, were composed of the last remnants of the soup.

One of the main casualties following the emergence of Neptunism was Diluvialism (Chapter 5), the idea that the biblical version of Noah's Flood produced all the observable rocks and fossils. Even though many of the ideas appear to be the same, the underlying philosophy was different. Many of the Neptunists did not believe in and would not support the concept of a universal flood in a biblical sense, even though they based their ideas on the presence of a universal ocean.

As Germany derived a significant proportion of its income from the mining of metals, the exploration and study of mineralogy and the formation of minerals were actively encouraged by the government, with individual states setting up their own schools of mining. It fact, it was thought that the entire Earth's crust was formed by precipitation and crystallization, in which water obviously played an essential role. In Germany, this formed the basis on which geologists, mining engineers, and miners could start to predict where particular minerals might be found (this was quite an advanced way of thinking at the time, as many others used guesswork or intuition to find their resources). By being able to identify the basic types of minerals and their associations, and establish how they were formed, they could use this information to locate mineral veins. This is a good example of applied rather than theoretical geology.

It was assumed that any rock that had a crystalline structure had been deposited in water and that "the age of rocks everywhere could be told from their composition". Granite Mountains appeared as the first deposits from the cloudy primeval ocean and therefore their presence provided the evidence that water had covered the entire Earth.

Werner, as we have already seen in Chapter 2, developed a five-fold classification of rocks, depending on whether the rocks were the original chemical precipitates or derived rocks. His ideas have frequently been presented as a simple "onion skin model", in which each layer built up on the last, firstly as a precipitated layer followed by additional, more localized layers that were formed due to subsequent erosion of the original rocks as the sea level gradually fell. However, he was well aware of the complexity of the geological formations in his local area and explained these as being a consequence of irregularities in the Earth's original shape. His primary rocks were precipitated directly onto the original surface and therefore followed its contours, and covered the entire surface. As the oceans subsided, erosion and further precipitation produced the other rocks. Sea level at the time of formation defined how high up the mountains these rocks could exist: the top of the mountains were therefore always formed of granite and the youngest rocks were always found on the lower slopes, in the valleys and plains. He also thought that settling of the different layers of rocks, after they had been formed, led to localized folding and faulting.

Discussion point

It is important to note that both of these ideas were based on relationships that Werner and his contemporaries had observed in the field. In other words, they were not simply theoretical, old fashioned, or fanciful ideas, but based on clearly recorded field relationships.

A gradual decrease in sea level in the Baltic appeared to provide field evidence for a continuous drop in sea level.

The strength and popularity of his theory lay in its ability to organize all minerals by identifying each formation with a particular period in the formation of the Earth's crust, effectively following the Principle of Superposition (Chapter 2). It meant that the theory could be used as a practical, predictive model – particularly for locating mineral resources, which of course was the main reason that Werner had developed it.

However, the idea of a retreating ocean led to a rigid framework of sequential events, i.e. granites and basalts, the original precipitates, could not have been formed at a late stage in the Earth's history, because they had to be formed during the original precipitation event. Werner recognized that, in practice, different types of rocks were associated together in certain formations that did not fit in with his model of a retreating ocean.

He even used fossils as a means of dating rocks: he recognized that fossilized remains in his oldest, transitional rocks were always of primitive life forms that, he thought, lived in an ocean containing a large proportion of dissolved chemicals. As precipitation continued, the ocean became cleaner and more advanced life forms developed. Eventually terrestrial life forms began to appear on the exposed land surface. This implies that there was a progression in the development of life, although he may not have thought of it as evolution as we now consider it (Chapter 6).

Werner was considered to be a master teacher, and people came from all over Europe to study with him. This meant that he had a tremendous influence on a large number of geologists who then followed his ideas.

Werner's ideas were based on a rigid Earth's crust. As his followers went further afield, from Siberia to the Alps, the evidence they collected indicated that subterranean explosions had thrown up the mountains from relatively shallow seas, as opposed to being precipitated in deep oceans. This showed that the basic principles of Neptunism needed modification. Werner was faced with an increasing volume of evidence that basalt was formed from a molten state. In response, he proposed that this was a result of localized melting due to chemical reactions. He quite rightly pointed out that modern volcanoes did not produce crystallized rocks such as granite.

Discussion point

Many of Werner's followers eventually changed their minds and began to see that, in the face of increasing evidence outside the area in which he lived, Neptunism did not work. However, it should be noted that it was this evidence that changed their minds and not the arguments of Hutton and the Plutonists.

It should also be remembered that, although Werner is generally regarded as the most prominent of the Neptunists, he was just one of many.

This leads to the following question. At which point do you continue to modify a theory to fit new data or abandon the theory and develop a new one? In other words, how far can you bend a theory before it breaks?

4.3 Plutonism

Plutonism is a theory based on the idea that heat from within the Earth generates granites and volcanic rocks and leads to uplift of the Earth's surface.

Hutton is usually portrayed as the forward-looking geologist, who led the campaign to discredit the retreating ocean idea, because he described the geological past in terms of what is happening today. To Hutton, volcanic rocks such as Arthur's Seat were an indication that the centre of the Earth was molten. Nevertheless, it is important to point out that he made no detailed studies of volcanic rocks and discovered his best evidence for significant Earth movements only after he had established the basic outlines of his theory (Chapter 5). This was "based on his own very individual vision of an eternally viable universe created by an all-wise God".

Hutton was also known for introducing the new farming techniques that he had learnt in Norfolk. During his time there, he studied examples of erosion along the coast at Great Yarmouth and recognized the link between sands and sandstones. It appears that it was from this and other observations, including his studies of erosion on his farm at Slighhouses and the formation of soil from eroded materials, that he reasoned that erosion of existing soil and rock was balanced by the deposition and generation of new soil. He began to think of the Earth as a self-repairing system whose purpose was to sustain life. This view was significantly different to all other ideas at the time, which centred on the Earth as being a steadily decaying system. Hutton viewed the Earth as a machine that would continue to operate through the processes of erosion, deposition, and uplift. He thought that as continents were worn down, others would rise from the sea in an endless cycle. As Baxter records, "with his series of great natural revolutions in the conditions of the Earth's surface, as Playfair described it, Hutton had made the world the mirror of the Newtonian sky". New mountains were elevated to balance those being worn down, in an eternal balance between uplift and erosion, which turned the world into a perpetual-motion machine.

One of the questions that remained unanswered before the publication of Hutton's work was how soil and sediments were consolidated to form rocks. As Baxter puts it: "There was no answer to be had in the conventional wisdom that prevailed in Britain and across the Continent".

As Jack Repcheck records in his book, *The Man who Found Time: James Hutton and the Discovery of the Earth's Antiquity*:

> An unusual event occurred in the summer of 1744 that may have had some effect on him. One day, a particularly violent storm caused a landslide near the top of Arthur's Seat … the landslide created a hollow, which today is called Gutted Haddie. Exposing a large piece of hard volcanic rock, the landslide demonstrated, in a most profound way, the power of erosion. It is almost certain that Hutton, being of curious mind, was one of the many residents of Edinburgh who went to observe the damage.

This may well have been one of the episodes that helped Hutton develop his appreciation of the processes of erosion and the importance of igneous intrusions. In 1755, his attention was drawn to the effects of heat by experiments undertaken by Joseph Black (1728–1799), the famous physicist and chemist, who had heated up samples of Magnesian Limestone and generated CO_2. Other experiments conducted in France had shown that pressure could change the effect of heat on rocks. Heat, Hutton thought, could provide the driving mechanism for his theory.

It was also his involvement in the founding of the Royal Society of Edinburgh in 1783 – which was "born out of jealousy, turf warfare, and bitterness" – when he was nearly 60 that prompted him to present his first paper. In 1785, he presented his ideas to the society in two parts. He then published the outlines of his theory in a 28-page abstract titled *Abstract of a Dissertation Read in the Royal Society of Edinburgh, Upon the Seventh of March, and Fourth of April 1785, concerning the System of the Earth, Its Duration, and Stability*. This contained all the essential elements of his famous 1788 publication titled *Theory of the Earth; or an Investigation of the Laws Observable in the Composition, Dissolution, and Restoration of Land upon the Globe*. As Repcheck relates:

> At the time of Hutton's explosive lectures, the announcement that the Earth was ancient was startling. It would be akin to being told today that the sun is not really the source of the Earth's heat and light or that there actually is complex life on the moon.

Although he farmed at Slighhouses for 10 years, the farm did not make much of a profit. Fortunately, he had already made sufficient money through his other activities, including his assistance in the manufacture of Sal Ammoniac (which was initially derived from camel dung), which allowed him to retire to Edinburgh at the age of 43 and be able to undertake his extensive travels. Although Hutton developed his ideas first and then sought evidence to support and develop them, he thought that first-hand observation of field examples was important – a process that has continued in geology, largely intact to this day. At this time, he had already undertaken the first of his grand tours of Britain to observe examples of erosion and similar processes, carefully noting down all of his observations. He set out to gather "specific information". As Baxter highlights:

> some modern commentators have tended to dismiss Hutton as an "armchair geologist", spinning elaborate theories from the smoky comfort of the Edinburgh clubs. But between 1750 and 1788 he would journey through nearly every part of Britain, except Cornwall, the Hebrides, and northwest Scotland.

Having presented his theory, he then set out to find more evidence to support it. In 1785, he found his first unconformity at Loch Ranza on the Isle of Arran; in fact, this was the first unconformity to be recognized in Britain. It is interesting to note that he found his second unconformity at Jedburgh in 1787 and his most famous one at Siccar Point in 1788, which was the year that he published his expanded version of the 1785 *Theory of the Earth* (Chapter 5). He had already observed an igneous dyke at Crieff in 1764 and had identified the igneous sill at Salisbury Crags on Arthur's Seat in Edinburgh (Fig. 4.3) in 1774. In 1785, he visited Glen Tilt, where he saw the granite contact with the surrounding country rocks. The following year he studied similar contacts and basaltic dykes in Galloway at Sandyhills Bay (Fig. 4.4); a year later, during a visit to the Isle of Arran, he observed the granite contact around the centre of the island.

Fig. 4.3 **Hutton's first unconformity at Loch Ranza, Isle of Arran (top left), Salisbury Crag (top right) and "Hutton's Rock" (lower left and right), Edinburgh, where the base of Salisbury Crag meets the underlying country rock**

Fig. 4.4 **Hutton's second, rather overgrown unconformity at Jedburgh (left), and the foreshore at Sandyhills on the Galloway coast (right)**

Siccar Point (Fig. 4.5) was not only the most striking of his unconformities; it also provided evidence for Hutton's cyclic ideas. The lower sequence of rocks were laid down as horizontal layers in water, and were subsequently compacted then folded into steeply dipping (tilted) layers of rock. Following this, they were subject to erosion that cut them down to an irregular surface over which later horizontal layers were deposited. Later, the whole sequence was again tilted at a shallow angle.

Fig. 4.5 **Two views of Hutton's most famous unconformity, Siccar Point, on the Berwickshire coast**

Discussion point

If there was ever a case for making a natural feature a World Heritage site, surely Siccar Point should be high up on the list of candidates.

As we saw in Chapter 2, 1788, the year that Hutton visited Siccar Point was also the year that Werner published his classification of rocks that, as Baxter says, "reasserted unequivocally his view that basalt has an aqueous origin". In 1791, Werner responded to his critics with the publication of his theory on the formation of mineral veins. It was his view that these had nothing to do with heat, but were formed in rocks below the seabed through the crystallization of minerals from injected ocean water above.

Hutton actually produced two versions of the work he did in 1788. The first was a full-length version of his 1785 paper, which was published in the *Transactions of the Royal Society of Edinburgh*. This generated a number of adverse and a couple of positive reviews. One was written by John Williams (1732–1795), a geologist and mineralogist who wrote a highly critical review in his book, *The Natural History of the Mineral Kingdom*. However, Richard Kirwan (1733–1812), the Irish scientist, wrote one of the most critical reviews. In 1793, he criticized Hutton's ideas in a 30-page review in the *Transactions of the Royal Irish Academy*. Repcheck quotes that the "very day after Mr Kirwan's paper was into his hands", Hutton "resolved to write a major expansion of the 1788 paper", which was eventually published in 1795. In fact, only two of the three planned volumes were completed before his death.

Kirwan and others had found ammonites (Fig. 4.6) apparently preserved in basalt at Portrush, Northern Ireland. In 1799, he used these as clear

Fig. 4.6 **Two photographs of ammonites preserved on the foreshore at Portrush, Northern Ireland**

evidence that so-called igneous rocks had been formed in water. Later, it was realized that the rocks were mudstones that had been baked by the overlying basalt.

In his system, Hutton thought that there were no primary rocks dating back to the origins of the planet, as proposed by Werner and others. As we shall see in the next chapter, Hutton thought that the Earth operated as a closed system in which everything was recycled. This meant that all rocks were repeatedly broken down and reconstituted. He thought that "the result, therefore, of this physical inquiry is that we find no vestige of a beginning, no prospect of an end". (We now know that the Earth is in fact an open system.)

We have also seen that Werner thought that volcanoes were all confined to recent activity, whereas the Vulcanists (people who believed that volcanoes played an important part in the formation of stratified rocks) opposed some of Hutton's views, as they thought that volcanoes had had a major impact on the evolution of the Earth. Their basic objection was to Hutton's view that igneous rocks originated from deep within the Earth – in other words, he was a Plutonist. At the time, it was generally thought that the Earth was gradually cooling down and shrinking, however the idea that it could have an internal heat that brought hot rocks to the surface implied that it would expand. One of the problems Hutton faced in justifying his ideas was that he did not have a source for the heat that his theory relied on to produce his continued cycles of activity and the formation of rocks.

It is recorded that Hutton's friend, the chemist James Hall (1761–1832), offered to conduct an experiment to prove that granite came from a molten material, but Hutton would not let him do it as he "opposed the procedure of seeking to discover Nature's grand design in the laboratory". He also thought that the pressure and heat within the Earth could not be reproduced in the laboratory. The experiment was not conducted until after his death, when Hall used samples of Whinstone melted in crucibles, which he then allowed to cool at different rates. He found that those that cooled rapidly turned to glass, but the ones that were allowed to cool for 12 hours or more looked similar to the original material. In a paper published in 1805, Hall highlighted the similarity between modern lava from Mount Etna and Whinstone. It seems odd that Hutton would avoid such experiments, given that it was the work of Black, Watts, etc. that provided such crucial information on which he could build his ideas.

Porter makes an extremely important point concerning this period of time. With the founding of the Geological Society of London and other regional societies, geology became an established science with separate and distinct boundaries from other sciences. Even so, the nature of its subject matter meant that it faced opposition. As Porter says:

> Geology accommodated itself to its external environment by drawing its public intellectual boundaries very narrowly. That narrowness was part of a bargain, which geologists in effect made with society. For the body of knowledge and working assumptions possessed by geologists were, after all, potentially radical and disturbing: a threat to contemporary common sense. The high antiquity of the Earth, its many revolutions, the impermanence of continents, the discovery of enormous fossil saurians, the confirmation of extinction, the general refrigeration of the earth – such ideas were genuinely not welcome to the Christian cosmology of the Napoleonic War years. To achieve a social licence to pursue such studies, geologists had to realize their science in the hardest, most factual, terms, and temporarily to refrain from explicitly projecting this science as part of a wider cosmology. Tracing the strata of England confirmed the real world, was patriotic and possibly even useful. Whereas to stress the theoretical aspects of geology was bound to be unsettling; was in a very literal sense, to take what had been the solid ground away from under people's feet.

It is worth bearing this in mind when reading the next chapter.

Further reading

The second half of Porter's book is well worth reading, particularly his description of Hutton's work, with regard to both Scottish and English geological ideas at the time.

Discussion point

The basic difference between Neptunism and Plutonism was therefore that the former accounted for the formation of crystalline rocks as being precipitates from a universal primeval ocean, whilst the latter considered that the cooling of molten material, which originated from deep within the Earth, formed them.

Many authors comment on the fact that Hutton's own book was largely ignored due to its length, extensive use of French and his "flowery descriptions". Most of his ideas only found their way into geological literature through his friends who stressed the science rather than his religious ideas, which were that everything was "designed to perfection by an all-wise God". This means that to a certain extent they effectively distorted his original views.

He proposed that the Earth acted as a closed system with a heat engine. He outlined much of the rock cycle, as we know it and use today; and established the foundations of uniformitarianism as an idea, giving us our first glimpse of the Earth's long geological history.

As Baxter and other authors note, at this point geology divided into two distinctly different camps, the Wernerians and the Huttonians. The Royal Society of Edinburgh, and the newly-founded Geological Society of London, favoured Hutton's ideas, but Werner had his supporters in Britain as well. Both groups could point to field examples, which supported their ideas, and as Baxter says:

> the stakes couldn't be higher. The whole of the future of the science was going to be shaped by the outcome of the debate, and a lot of careers and reputations were going to be made and destroyed.

Many areas of science still follow the same pattern of behaviour.

We have seen that the latter half of the 18th century saw the development of two opposing theories for the formation of the Earth's surface. One is portrayed as being forward-looking and progressive and the other retrogressive and based on biblical ideas. Is this strictly true or a rewrite of history by the "winning side"?

5
Uniformitarianism versus Catastrophism

5.1 Introduction

William Whewell (1794–1866) – a scientist, priest, philosopher, and historian of science – reviewed volume two of Lyell's *Principles of Geology*. As a result, he was the first person to introduce the terms "uniformitarianism" and "catastrophism" for the two sets of opposing views of geological processes. But, what is the difference?

Uniformitarianism assumes that the same natural processes that operate now have always operated in the past, and at the same rates within the same physical laws. Catastrophism is the idea that the Earth has been affected by sudden, short-lived, violent events, some of which have occurred on a worldwide scale.

Most geology textbooks still present the realization that the Earth had a history that extended over billions rather than thousands of years, as a triumph of careful observation that defeated the constraints of religion and superstition. They sometimes portray the difference between the two views thus; uniformitarianism triumphed because it provided a general theory that was at once logical and seemingly "scientific". Catastrophism became a joke and no geologist would dare postulate anything that might be termed "catastrophic" for fear of being laughed at. Later in this chapter, you will see that it is only in the relatively recent past that we have been able to move beyond this situation.

Discussion point

Gould makes the following observation: "like so many tales in the heroic mode, the standard account of the discovery of deep time [a long geological time scale] is about equally long on inspiration and short on accuracy".

Time Matters: Geology's Legacy to Scientific Thought, 1st edition. By Michael Leddra.
Published 2010 by Blackwell Publishing Ltd.

5.2 Catastrophism

Geological thought throughout the Middle Ages was dominated by biblical tradition. This was based on Genesis and in particular the story of Noah's Universal Flood. During the 15th and 16th centuries, this began to change as European explorers brought back news of previously unknown areas of the world, but the influence of the Genesis Flood continued.

Background

To understand why it dominated so many ideas in geology, we should clearly identify the basics of the flood story:

1. It lasted for 378 days (some sources quote 371 or 376 days).
2. There were initially 40 days of rain, after which the water continued to rise for another 110 days.
3. This was followed by 74 days during which the water was "going and decreasing" and the tops of the mountains were uncovered.
4. It was 40 days before the raven was sent out and another seven days before the dove went out.
5. There were another seven days before a second dove was sent and a further seven days before the third dove was sent.
6. 29 days later, the surface of the Earth was dry, but the ground was not completely dry until 57 days after that, when Noah removed the cover from the Ark.

Most catastrophist and creation science books highlight the following as a link between the Flood and the basic order of strata as seen in the geological record:

1. A large proportion of the fossil record is associated with life, deposition, and preservation in water.
2. The lowest and therefore the oldest rocks contain marine sediments, particularly deep-sea sediments. The sea that existed prior to the flood contained the most primitive marine life forms and therefore equate to the geological sequences from the Cambrian to the beginning of the Devonian in geological terms.
3. Many of the books say that hydrodynamic selectivity ensured that the more streamlined denser animals would be deposited first, i.e. trilobites and brachiopods, which are found in the lowermost stratigraphic horizons.

4. As sea levels rose, other more advanced marine-based life died and are preserved as fossils in overlying rocks.
5. Due to the selective nature of deposition in water, there are concentrations of particular animals at particular levels, hence we find index fossils (specific fossils that can be used to relatively date specific rocks sequences).
6. As the water level continued to increase, more mobile "advanced" animal life forms were able to stay out of trouble. This, they say, accounts for the higher life forms such as vertebrates with pelagic (open ocean) habits, such as the Devonian fish "graveyards" in sandstones and the coal measures, which were brought down by torrential streams and rivers into the sea.
7. With the continuing rise of sea level, even higher-order animals were trapped, such as amphibians, reptiles, and finally mammals, depending on local circumstances, hence these are only found in rocks near the top of a geological sequence.
8. The above accounts for the apparent development of life as seen in the fossil record.

Discussion point

What are your views on the following questions?
Does the above represent the observed overall geological succession of both rocks and fossils?
If so, is this on a general scale or does it fit at a detailed, large-scale level of observation?
Do the above provide an adequate account for marine fossils found on mountains?

Although it is true, the majority of sediments contained in the rock record were deposited in water and the geological successions, in general terms, follow the trend outlined above. This is a gross simplification. As we saw in Chapter 3, there is extremely good geological evidence, which indicates that life did not come on to the Earth's land surface until the Devonian Period. It is therefore hardly surprising that only water-based fossils can be found in pre-Devonian rocks.

The majority of the sedimentary rocks we find do not show signs of rapid, large-scale deposition in water that would be consistent with deposition during a large-scale flood. Equally, deposition would have to have been rapid, as the effects of the flood only lasted for 371 days. During this period, all the fossils and all the rocks containing them would have had to be deposited.

If all the rocks were deposited in, and by water, how is it possible to account for the presence of desert sandstones, evaporite sequences, undisturbed reefs and lagoons, deep marine shales, and even coal measures with rootlets and trees still in place in the fossil soils, and even ancient glacial deposit in the geological succession?

Details of the rock and fossil record – even if it is incomplete (see comments below) – do not support such sweeping generalities. Even the selective use of "science" in the form of hydrodynamics is misleading. Having spent many years conducting experiments to study the engineering properties of sedimentary rocks, it is obvious that sedimentary rocks take a significant time to compact, dewater, and lithify. Waterlogged sediments have very characteristic features, which we do not find in the vast majority of rocks.

As explorers such as Columbus brought back animals, plants, and stories of other peoples from their travels, the premise behind the Genesis Flood came under increasing pressure. It led to observations similar to those listed below:

1. It was realized that the world was significantly larger than had previously been thought.
2. Explorers encountered indigenous people wherever they landed, even on remote islands far from known continents.
3. Huge numbers of new species of both plants and animals were found that were previously unknown in Europe, Africa, and Asia.

Each of these discoveries challenged the idea of a universal flood, leading to questions such as:

1. How could water from an unknown source cover the entire globe?
2. How could so many different people, living in the far-flung corners of the globe, often separated by large oceans, have descended from the survivors on the Ark?
3. How could the huge variety of animal species then known, each of which lived in specialized niches, have travelled to the Ark prior to the flood and travelled back from the Ark following the flood. Especially considering that they would have had to travel through what would have been, to them, generally hostile environments (such as over the land or across vast oceans), when the majority of them would be unable to make such journeys?
4. How could the Ark hold all the new species that were being discovered?

5. How could Ussher's and similar time scales be true when explorers had encountered peoples with significantly longer recorded histories?

Discussion point

As Alan Cutter says, "ever since the discovery of the Americas, with their abundance of previously unknown animal species, biblical literalists had been in a tight spot. How could all of these species have fitted on Noah's Ark?"

This also raised questions about why some species on other continents could be different from their European counterparts. Interbreeding following the Flood was proposed as one of the answers, but naturally, the non-European species were considered as "degenerate versions".

Problems like these still remain. In fact, the more we find out, the more problems result. In his book, *A Biblical Case for an Old Earth*, which was published in 2006, David Snoke makes the following points with regard to the scientific problems of flood geology. Firstly, he says that the most popular description of the depth of water would require a "six-mile-deep flood covering the entire globe". He then lists a number of factors requiring miraculous interventions to fulfil the flood story. These are:

1. The transportation of millions of animals from Australia, the Americas, Antarctica, and all the islands across the world's oceans to the Ark. These included many species of animals that live in small ecological niches far enough away from the Middle East, to make it impossible for them to reach the Ark.
2. Given the dimensions of the Ark as specified in the Bible, how did they all fit in? In addition, he questions how the Ark could have held the volumes of specialized food and fresh water required to keep such a diverse range of animals alive for their time at sea.
3. There were just eight people looking after the millions of animals on board the Ark. How would they have had the time and resources to dispose of their faecal waste each day?
4. The occupants would have had to withstand soaring heat in a confined space of the Ark during the voyage. He quotes temperatures of "hundreds of degrees" generated by the close proximity of all the animals in that "windowless box". These conditions would also preclude the provision of special climatic habitats for its occupants.

5. Snoke observes that the Earth's water moves from one place to another without significant increases in overall volume. He therefore questions where all the extra water that appeared "out of nowhere and the destruction of that water afterwards", came from. Some of the ideas that have been put forward include:

   a. That water was stored in large areas underground; the "springs of the deep", from which it flowed to the Earth's surface and then flowed back again.
   b. That "the water resided in a cloud 'canopy', but how could the atmosphere hold enough water to cover the Earth to a six-mile depth"?

   He discounts the first of these by questioning where the volume of water, required to flood the entire surface of the Earth to a depth of up to six miles, could be stored. For the second, he observes that such a cloud would have made the Earth's surface hotter than that of Venus. This blanket of cloud would also have reduced the amount of sunlight reaching the planet's surface, resulting in almost, if not complete darkness during the event – something that does not appear in the description of conditions during the Flood as presented in Genesis, chapters 7 and 8.
6. The weight of the additional water on the land surface, would have caused the Earth's continents to sink, but this would not have been significant over such a short period of time. (See Chapter 8 for an explanation of isostatic adjustment.)
7. The trees and plants submerged in the water would have died. As Snoke quite rightly points out, in Genesis 8:11, the dove that is sent out from the Ark arrives back with a fresh olive leaf shortly after the waters subside. This suggests either that the olive tree survived underwater or that the land surface was not completely covered. If neither were true, it also begs the question, how could the tree have grown back so quickly?
8. It is not unusual to find trees preserved in their life positions in coal mines and outcrops, such as the one shown in Fig. 5.1. Coal seams have sometimes been used to explain that forests were swept away and deposited in specific layers by the flood. However, if you look at most coal seams, you can still see the tree root systems preserved, in life position, within the fossil soils (termed a Seat Earth). How could these have survived, having been subject to such a dramatic event? (See Chapter 8 for a further discussion on an alternative theory for the generation of coal.)

Fig. 5.1 **The famous fossil tree preserved in the village of Stanhope, County Durham, which was recovered from local Carboniferous rocks**

Discussion point

In response to flood scientists' criticisms that modern geologists will not accept evidence of a global flood, because they have to uphold existing, traditional ideas, Snoke adds that:

> some people claim that geologists are too unwilling to change their beliefs from traditional theories. Actually, in the past forty years, geologists have accepted two major, new theories in response to new discoveries.

The first of these was Continental Drift/Plate Tectonics, which, as we will see in Chapter 8, took some time to be accepted until there was sufficient evidence to "prove" it. This was especially so in America. The second is the theory that a large meteor struck the Earth, which led to a major extinction event that included the demise of the dinosaurs. Evidence for this was gradually collected from across the globe until there was sufficient information to show that such an event had occurred. Some people, including those who were looking for the evidence, were for some time reluctant to accept the idea (Alveraz 1997). As Snoke says, "In each of these cases, revisions in the theories of the entire history of the world were supported by substantial, global evidence". He concludes this with the following comment – "geologists accepted these new theories because they feel the evidence warranted a change of mind. No similar evidence has come up that has convinced geologists of a global flood".

Further reading

It is worth reading books such as *Noah's Flood: The New Scientific Discoveries About the Event that Changed History*, written by William Ryan and Walter Pitman. This book details the scientific investigations that appear to show that the flooding of the Black Sea 7,500 years ago may provide the basis for the many flood stories that exist in the Near and Middle East, including Noah's Flood.

5.3 Diluvialisim

Towards the end of the 17th century, the literal reading of Genesis moved away from merely thinking about how all of the animals could fit in the Ark and which ones had been taken on board, to start to consider the geological effects of the flood. This led to a school of thought that is frequently referred to as "Diluvialism". This name was derived from a term *Diluvium*, which was introduced by William Conybeare in 1822: he used this for the water-worn debris he identified as originating from the effects of the last great universal catastrophe (Noah's Flood). Remember, Conybeare was partly responsible for the establishment of the Carboniferous and Cretaceous successions (Chapter 3), so he cannot be dismissed as a non-entity or someone who did not understand the nature of rocks, how they were formed, or the order in which they were laid down. As we shall see in Chapter 6, he also helped Mary Anning with her fossil reconstructions and descriptions and is said to have taught Buckland and Sedgwick.

One of the most important and influential Diluvialists was the Reverend Thomas Burnet (1635–1715) (Fig. 5.2), who was Royal Chaplin to King William III. He is usually taken as being one of the classic catastrophists and the standard view presented in many textbooks is that publication in 1691 of his book, *The Sacred Theory of the Earth*, showed that he produced his theory for the formation of the Earth purely based on strict reading and interpretation of the Bible, and that his theories therefore slowed down the progress of science. One description of his book (in the *Nuttall Encyclopaedia of General Knowledge* published in 1907) included the phrase, "descriptive in parts, but written wholly in ignorance of the facts". Nevertheless, Burnet could be considered as one of the Britons who initiated the process of developing theories for the Earth's formation and history. His book has also been described as "the most popular geologic work of the 17th century". One interesting analysis of Burnet's work is presented in Sir Archibald

Fig. 5.2 **Reverend Thomas Burnet**

Geikie's *Founders in Geology*, which was published in 1897. In this he describes Burnet's theories as "monstrous doctrines" and also states, quite rightly, about Burnet and other people who held similar theories that:

> it was a long time before men came to understand that any true theory of the Earth must rest upon evidence furnished by the globe itself, and that no such theory could properly be framed until a large amount of evidence had been gathered together.

It is important to note that Geikie talks about how important it is to base theories on observable evidence, a practice that did not come into force until the 18th century. From this, it would appear that Burnet was considered a poor scientist. However, Newton commented that his theory "was an exemplary representative of a scholarly style valued in his own day". (Note that the term "science" itself did not exist as such until the 19th century.) He also thought that Burnet's theory was quite sound and felt that it was more probable than his own ideas.

Burnet's ideas were based, as were most other people's at the time, on his belief that the Bible was, "unerringly true". He looked for scientific explanations for the biblical account. It is important to understand that he did not believe in divine intervention as an answer to the problems that could not be solved by established physics. Burnet's basic argument was that God got it right first time: he ordained the laws of Nature to yield an appropriate history, and therefore did not need to intervene later to patch things up in an imperfect universe by miraculous alterations to his own physical laws.

Burnet viewed the Flood (like most other catastrophists) as crucial to his ideas. He calculated the average depth and extent of water in the oceans (which he underestimated), to determine whether the water available could bury the continents. He determined that 40 days and nights of rain would add little to the total amount of water on the Earth's surface and that, as no additional water could be added by divine intervention, there had to be a worldwide layer of water below the Earth's crust.

However, how could a layer of water exist below the Earth's surface? According to David Standish's book, *Hollow Earth*, Burnet thought that "from the original liquid chaos things settled out according to their densities, the heaviest forming the core, with the 'liquors' of the earth rising towards the top". Burnet thought that the centre of the Earth was largely composed of water and cited as evidence for this the existence of water-filled caves and lakes. As the water was released, the crust cracked, giving the appearance of a "hideous ruin" or "a broken and confused heap of bodies". At the time, many people thought that before the Flood, the Earth had been a perfectly smooth sphere. Burnet also thought that the cracking of the crust led to the tilt in the Earth's axis, which introduced a seasonal climate. Obviously, this implies that before this event, the Earth's climate was non-seasonal. He then used this basic premise to work backwards to pre-flood times to reconstruct what he thought the surface of the Earth would have looked like. As for the beginning, he argued that the Earth did not rotate at a slower speed (as suggested by Newton) but, as the Sun was not created until day four, the length of the first four days would not have to be the same as at present. This would therefore allow for an extended history rather than four literal days, for the formation of the Earth.

It is important to note that Burnet believed that the Earth had a definite history, which implies a change over time. He also thought that its present surface was not in accordance with the original design as "the mountains are continuously destroyed and washed into the sea and nothing is brought back" – a view that was typical of the widely held belief that the Earth was in gradual decay.

Although most people considered *The Sacred Theory of the Earth* to be a masterpiece, Burnet was criticized at the time by the Church for not paying enough attention to the details of the biblical text. He was also criticized by others such as Hooke for not taking into account fossils and rock strata. Hooke felt that the time period over which the flood had existed was insufficient to account for the features Burnet was ascribing to it. It is also interesting to note that Burnet thought that the Genesis version of the formation of the Earth was just a story and because of this, according to Standish, "King William took considerable heat over having someone with such scandalous views as his court Chaplin and had Burnet fired".

Another critic of Burnet was John Woodward (1665–1728). He was a leading naturalist and Professor of Physics at Gresham College. Through his work and contacts, not only in Britain but also throughout the world, Woodward had a clearer understanding of the nature of the Earth's structure and the relevance of fossils. However, he was more concerned with the effects of the Flood rather than the mechanics by which it happened: he maintained that as the Flood subsided, heavier particles would settle out first and lighter ones would settle on top of these, which is a repeatable process seen in every river, lake, sea, sand, ocean, and desert today. Thus, the sequences of rocks should show a decrease in density upwards – that is, all the older rocks should comprise coarse sandstones and pebble beds and all younger rocks should be fine-grained clays and mudstones (think carefully about this concept). He thought that "all minerals, stone and marble; all metals; mineral concretions and fossils" were "borne up in the water and was again precipitated and subsided towards the bottom", depending on their "quantity or degree of gravity". Woodward considered that the Flood was caused by a temporary "shutdown" of Newtonian gravity (he considered Newton his only intellectual equal), and faced as much criticism for his views as Burnet had for his. Unfortunately, field evidence did not support Woodward's ideas; rock sequences did not occur in the order predicted by his model.

Discussion point

As Cutter reports, "Woodward proposed that such an event, decreed by God, would cause all of the Earth's solid matter to 'instantly shiver into millions of Atoms and relapse into its primitive 'Confusion'". Following the Flood, he thought that gravity would cause the heaviest particles to settle out first and this would account for fossils being found deep within the Earth. Many people at the time referred to this as Woodward's "hasty pudding" theory.

Burnet was not the only person who thought that the water required for a Universal Flood existed below the Earth's surface. Edmund Halley (Chapter 1) used both mathematical and physical principles to support his claim that the shock of a comet passing close to the Earth could have caused the oceans to overflow and the Earth's axis to tilt. He claimed that the drag caused by the movement of such vast amounts of sediment-charged water could have led to the formation of the mountains. Interestingly, although he had these ideas, he delayed publishing them for nearly 30 years.

Following his own recordings in the Atlantic and Pacific oceans, together with the study of over 100 years of magnetic data, like Burnet, Halley also thought that the Earth consisted of a series layers. He proposed the theory that it was made up of a number spheres, surrounded by a fluid, each of which had its own magnetic poles, thus accounting for the observed drift in magnetic data. Standish quotes Halley as saying, "something is moving down there to cause this shift". He presented his ideas for a hollow Earth to the Royal Society in 1692. As Standish says, "his theory of a hollow earth was the first scientific hypothesis to draw on Newton's ideas, and it wasn't as off the wall as it may seem". In fact, we now think that the Earth is effectively made up of a series of moving spheres, some of which generate the Earth's magnetic field (Chapter 2).

The idea that the Earth contained water was not new. Athanasius Kircher (1601–1680) – a German Jesuit who was "the last man to know everything" – published his *Mundus Subterraneus* in 1664. Kircher had visited a number of volcanoes in Italy and thought that the Earth contained a series of "pockets of fire" that fed the volcanoes and heated up subterranean water, which fed via a "great vortex" at the North Pole into the Earth's interior. From here, it was then heated and released as warm water at the South Pole.

In 1878, Americus Symmes also proposed the idea of a hollow Earth composed of a series of spheres with a large opening at the North Pole.

Further reading

For a good review of these and other Hollow Earth theories, it is worth reading the first two chapters of David Standish's book, *Hollow Earth: The Long and Curious History of Imagining Strange Lands, Fantastical Creatures, Advanced Civilizations, and Marvellous Machines below the Earth's Surface.*

Halley's, Burnet's, and Newton's ideas were used by William Whiston (1667–1752), an English theologian, historian, and mathematician who succeeded Newton as Lucasian Professor of Mathematics at Cambridge University. He developed his text, *A New Theory of the Earth from its Original to the Consummation of All Things,* which was published in 1696. In it, Whiston proposed that the Earth was formed by gravitational differentiation of material within the tail of a comet. The Earth's axis was then tilted as it passed through the tail of another comet for between 10 and 12 hours on Thursday 27 November (no year given). The comet also distorted the shape of the Earth, changing it from a sphere to an ellipse, a process that released a vast quantity of water from a subterranean source, leading to Noah's Flood. As the water subsided, rock strata were deposited that contained the fossil remains of animals alive at the time. This led to major objections from conservatives who thought he was taking liberties with Genesis, but his comet idea was similar to that of Edmund Halley.

Georges Cuvier, Professor of Zoology at the College de France, studied Cretaceous, Tertiary, and Quaternary rocks exposed in the Paris Basin, where he noted that in the topographically flattest parts of the basin, the strata were generally flat-lying, but around the foothills of mountains, the underlying sediments were inclined and deformed. This he thought implied that the rocks were of two different ages. He believed that the last "great revolution" (flood) was witnessed by humans but that it was not a unique event; rather it was just one of many such events.

Professor Robert Jameson (1774–1854) (Fig. 5.3), a naturalist and geologist at the University of Edinburgh, produced one of the first translations of Cuvier's original manuscript in 1817. In so doing, he distorted its emphasis to provide a closer match to his own views.

It is said that one of Jameson's motives for doing so was to attack Cuvier's teachings as being contrary to that of Werner and Moses, who

Fig. 5.3 **Professor Robert Jameson**

both talked of the Universal Flood as a unique event. Jameson attacked Cuvier on three counts:

1. He thought that the steeply inclined strata Cuvier had observed in high mountains were original and not the result of crustal "dislocations".
2. That erratic blocks found in the Jura, that formed part of Cuvier's evidence, were not the result of eruptions from their source but were transported by water, i.e. the Flood. It was generally thought that erratics, large blocks of rock that are not associated with the local vicinity, could only have been transported by large-scale floodwaters.
3. To Cuvier's "thousands of ages", Jameson replied, "our continents are not of a more remote antiquity than has been assigned to them by the sacred historian in the book of Genesis, from the era of the deluge". In other words, they were not extremely old but only as old as that calculated by Ussher and others using the Bible as a reference.

By the middle of the 18th century, ideas based on the concept of rock formations and the distribution of fossils being due to a universal flood were beginning to recede. This change had been brought about by extensive fieldwork throughout Britain and Europe, which indicated that the majority of rock sequences studied showed little evidence of catastrophic fluvial deposition. Fossil and stratigraphic evidence indicated that they were associated with specific environments rather than having been deposited by a universal flood. Even though the idea that the Genesis Flood deposited all rocks was waning, many key geologists still used the concept of a universal flood as the basis for the deposition of superficial rocks and sediments.

For example, George Bellas Greenough (1778–1855), a student of Werner's and a Member of Parliament and one of the founders and the first President of the Geological Society (in 1807), published *A Critical Examination of the First Principles of Geology* in 1819. In this he said, "that if seas, rivers or collapsing lakes could not transport the exotic blocks found all over Europe then only the existence of a universal deluge could account for them". He proposed that the trigger for the flood was the collision of a meteorite with the Earth.

Other important and influential geologists were also confirmed Diluvialists. These included William Buckland (1784–1856) (Fig. 5.4) – Professor of Geology at Oxford, a Minister in the Church of England (and later the Dean of Westminster), President of the Geological Society from 1824–1825 and 1840–1841, and an inspirational lecturer. He tried to reconcile the biblical version of creation with scientific theory. He thought

Fig. 5.4 **Professor William Buckland**

Fig. 5.5 **The entrance to Buckland's Cave at Kirkdale, North Yorkshire**

that the essential construction of the Earth occurred over a long period of time before the events recorded in the Bible. He used a bone-filled cave at Kirkdale, North Yorkshire (Fig. 5.5) as evidence of a universal flood, although he never actually said that his flood was the same as Noah's Flood.

The cave had been discovered by quarrymen in 1821, who contacted Buckland so that he could have a look at what they had found. During his excursions into the cave, he found bones belonging to bears, bison, reindeer, elephants, hippopotamuses, rhinoceros, and over 300 hyenas.

Discussion point

The question Buckland faced was, how could such a wide range of animals get into a cave when the entrance was only 1 m high (Fig. 5.5). In addition, why were hyenas, elephants, and reindeer together in the North of England, miles away from their natural habitats?

At this point Buckland showed his scientific credentials. He decided to try to find out how the bones could have got into the cave, and imported a hyena into his laboratory at Oxford to study how it behaved. He noted that it dismembered its prey before dragging it off to eat, and identified this as the method by which the larger animals found their way into the cave. However, he did not do this until after he had presented his ideas to the Royal Society. His methodology and interpretation of the bones and sediments in which they lay, and the deposits above and below, has been described as an "exemplar" of scientific work. Humphry Davy, President of the Royal Society at the time, emphasized the importance of Buckland's work as it provided a fixed point in geological history, which appeared to coincide with similar finds across Europe.

As for the question of how such a diverse range of animals could have collected in the same area, Buckland partially reverted to the Bible. Buckland's interpretation, as usually reported, is that as the floodwaters rose, the animals would have migrated in front of them. When they drowned, their bodies floated into this area of Yorkshire, where the hyenas dragged them out of the water, ripped them apart, and took the bits into the cave to eat. As the floodwaters continued to rise, the hyenas were also trapped and died in the cave. However, Rudwick indicates that Buckland thought that all the animals actually lived near the cave, i.e. they lived together, at the time in North Yorkshire, an interpretation that has clear environmental implications.

Discussion point

We may think that the idea of a universal flood is crazy, and that Buckland's solution based on the evidence he found seems flawed. However, the cave, which is situated on the southern edge of the North Yorkshire Moors, is surrounded by large valleys that today contain only small, misfit streams (Fig. 5.6). We now know that these were the result of glacial outwash, but Buckland would have been unaware of this at the time.

In the early 19th century, it was widely believed, based on field evidence, that river valleys were formed by prehistoric, major flood/tsunami events. Again, this may seem to be an extreme interpretation of field evidence, but Fig. 5.7 shows a series of valleys in Northumberland, in which water clearly flowed "uphill" in some places.

Fig. 5.6 **A typical valley on the North Yorkshire Moors close to Buckland's Cave at Kirkdale**

Fig. 5.7 **A series of dry valleys in the Cheviot Hills, Northumberland**

Two years later, in 1823, Buckland was shown a cave on the Gower Peninsula in South Wales: this cave was significantly larger than the one in Yorkshire. Among other fossils found in the cave, was the skull of a mammoth and the entire left side of a fossilized human female skeleton together with ivory ornaments. This presented a problem: firstly, it was believed at the time that humans appeared on the Earth after animals such as the mammoth became extinct. Therefore, how could a fossilized human possibly be found with that of a mammoth? Buckland decided that the layers of sediment in the cave must have been disturbed to allow them to be found together. Both the human remains and the ivory ornaments were coated in red minerals, which led to the legend of the Red Witch or the Red Lady of Paviland.

Discussion point

In both cases, Buckland used good scientific methodology to establish the occurrence of the bones, but took a "leap of faith" to provide an interpretation with regard to the presence of exotic animals.

Having taken field trips to the Kirkdale Cave for many years, it was easy for students to come up with feasible explanations for the bones. However, when they were restricted to the evidence and ideas that would have been known when Buckland visited the cave, their explanations proved a more difficult task. Without being able to use the ideas of climate change, glaciations, and changes in sea level, etc., the students almost invariably had to resort to the same conclusions as Buckland.

The problem of trying to remember what people in history would or would not have known is a difficult one. Sometimes it is very easy for us to look back at their views and forget that they did not have the information, ideas, or concepts that we take for granted today.

Other people thought that the idea of Britain being inhabited by such animals that clearly lived in a very different climate was ridiculous. Deborah Cadbury reports in her book, *The Dinosaur Hunters*, that George Fairholme (1789–1846), described as a scriptural geologist, who published *A General View of the Geology of Scripture* in 1833, thought that "had this not been the hypothesis of some of our ablest geologists it would have been termed the result of the most inconsiderable ignorance". At the time, Buckland was widely regarded across Europe as the scientific "heavyweight".

Christopher McGowan, author of *The Dragon Seekers: the Discovery of Dinosaurs during the Prelude to Darwin*, points out that Buckland's "denial of human species being contemporaneous with extinct ones had ramifications for other investigators". He records that a Reverend John MacEnery found flint tools beneath a thick layer of stalagmites in a cave in Devon. This indicated that the flints were extremely old and must have predated the formation of the stalagmites; however, MacEnery was prevented from publishing this "until after Buckland's death". Even then, when the results of a subsequent investigation were published, they did not refer to the flint tools. Palmer observes that "Buckland's sincere attempt to reconcile the Old Testament account with the merging geological facts further delayed the recognition of the scientific facts of the matter. However, Buckland was not a dogmatic fundamentalist".

Buckland's conviction that a universal flood existed and had been the source for all rock deposition, began to wane as further evidence increased, particularly with regard to the formation of volcanoes.

Another crucial line of evidence, which convinced many geologists at the time that exotic blocks – erratics – had not been deposited by water, came from the study of the effects of glaciations. At the start of the 19th century, Hutton and Playfair described the effects of glaciations but their ideas were largely ignored. However, other Earth scientists working on the continent, such as Horace Saussure and Agassiz, were also investigating the effects of glaciers.

Louis Agassiz (1807–1873) (Fig. 5.8), was a Swiss-American naturalist, who had been a student of Cuvier before becoming the Professor of Natural History at the University of Neuchatel; he later became an expert on fossil fish. In 1837, following extensive fieldwork in Europe with his friend Jean de Charpentier (1786–1855), a German-Swiss geologist, he became convinced about the possibility and effects of glaciations. According to William Ryan and Walter Pitman's book, *Noah's Flood: New Scientific Discoveries*

Fig. 5.8 **Professor Louis Agassiz**

About the Event that Changed History, Agassiz, in a lecture to the Swiss Society of Natural History at Neuchatel:

> … chose quite impulsively to abandon his prepared lecture on Brazilian fossil fish … Instead he publicly confessed an almost overnight conversion to a new theory that a vast 'ocean of ice' had once covered the whole surface of Europe and all of northern Asia as far as the Caspian Sea.

They add that "although Agassiz dropped his theory of a prior glacial epoch like a bombshell, it fell on closed ears".

Agassiz attributed the onset of continental glaciers to the last mass extinction (see later in this chapter for a discussion of extinctions). Although he was not sure of the actual cause of the ice age, he felt that the heat loss from dead bodies involved in an extinction would lead to a drop in the surface temperature of the Earth to freezing point. He argued that, as a result of the "greatest catastrophe which has ever modified the face of the Earth", the Alps were thrust up so that "huge blocks of rock were propelled skyward, and landing on the up-arched ice, slid great distances in all directions" to form erratics blocks. Following the return of organic life, the associated body heat would warm up the Earth's surface again, and the glaciers returned to their present position.

According to Ryan and Pitman, Buckland – who was "the most important member of the opposing establishment" – remained unconvinced that glaciation could be the key to the distribution of the exotic blocks until he went on a field trip with Agassiz in 1840. This trip allowed him to look at the "drift" deposits at Blackford Hill, south of Edinburgh. Having looked at a sequence that Buckland thought clearly showed that they must have been deposited by the flood and could not have been produced by glacial activity, Agassiz is said to have presented him with an opposing view. He took him to a nearby cliff, which showed a series of striations (caused by ice grinding rocks against its surface), at which "Buckland's conversion from Diluvialism was instantaneous". Other authors say that it was during field trips with Agassiz in the Jura Mountains in 1838 and 1840 that Buckland became the first British geologist to accept the role of glaciers in the development of the landscape. He then related the features he saw in the Jura Mountains to the ones he had observed in Scotland. Either way, Buckland gave up his Diluvialist ideas and became a glaciologist. Lyell, who had also insisted that erratics were the work of the Flood, eventually followed suit, but it is said that he argued with Agassiz "for decades" about the possibilities of ice ages as the idea conflicted with his views of uniformitarianism (see below).

Another famous Diluvialist was Adam Sedgwick (Chapter 3), who in 1831 abandoned the idea and suggested that many geologists had been too hasty in assigning the deposition of all superficial materials to the work of a universal flood. George Greenough, having been convinced by Charles Lyell (who visited Auvergne in France with Roderick Murchison) also rejected the idea of a universal flood as the mechanism by which superficial deposits had been laid down. Hutton also thought that erratics had been moved by ice, whilst Sedgwick thought they had rolled down the sides of mountains as they were being uplifted.

Eventually Buckland, Agassiz, and Lyell gave important lectures on the effects of glaciation to the Geological Society.

Discussion point

Does the above fit with the view of Buckland being a "blinkered" geologist who is tied to the Bible, stuck in the past, rigidly upholding traditional views?

As Rudwick writes, far from:

> ... retarding the Progress of Science, a lively concern to understand Genesis in scientific terms, and more particularly an interest in identifying the physical traces of the flood, facilitated just the kind of thinking that was needed in order to develop a distinctive *geohistorical* [his italics] practice within the science of the earth.

5.4 Uniformitarianism

As we have seen, James Hutton's *Theory of the Earth*, first published in 1788, is usually taken as the point at which an extended Earth history was proposed in Britain (Fig. 5.9). As noted in Chapter 4, Hutton's book was generally regarded as being virtually unreadable because it had pages of text consisting of pure quotes in French, and its "flowery language" made it difficult to follow. Rudwick makes the point that during the period in which Hutton lived, if anyone wanted to publish their work to make it available to the wider scientific community, it had to be written in French (the language of science). Therefore, Hutton's use of French, far from being unusual, would have been the norm. It is generally reported that it was John Playfair's rewrite that made it comprehensible, and which opened it up to be more widely read (Fig. 5.10). Yet in rewriting it, Playfair made changes to some of Hutton's original concepts.

The final two lines of Hutton's original 1788 book contain one of the most quoted lines in geology:

Fig. 5.9 **James Hutton**

Fig. 5.10 **John Playfair**

> If the succession of worlds is established in the system of Nature, it is vain to look for anything higher in the origin of the Earth. The result, therefore, of our present enquiry is that we find no vestige of a beginning – no prospect of an end.

Interestingly, Repcheck reports that during his lecture to the Royal Society of Edinburgh, Hutton referring to the processes of erosion said, "with respect to human observation, this world has neither a beginning nor an end". Repcheck notes that "Hutton was arguing that the earth is unknowably old, not eternal; the phrase 'with respect to human observation' is crucial in this context".

As we saw in Chapter 4, the traditional view is that it was Hutton's observations and ideas on unconformities and igneous boundaries, and the Earth's mechanisms for renewal, that were revolutionary. Hutton thought that unconformities showed that the Earth was capable of repairing itself through uplift, subsidence, and deposition. It therefore went through an unlimited series of cycles of renewal and decay (Chapter 4). The important implication of this idea was that for the first time, geological processes did not impose a time limit on geological history, as had previous theories. Hutton viewed the Earth as a machine that worked in a way that prevented ageing. He also felt that how the "machine" began was beyond the realms of science, hence "no vestige of a beginning". Once it had started, it could not stop of its own accord and therefore there was "no prospect of an end". Each cycle worked in three stages:

1. Rocks were eroded and transported into the oceans as sediments.
2. The sediments were deposited as horizontal strata, which as their layers built up produced heat and pressure.
3. This heat and pressure melted the sediments that were then intruded as magmas (molten rock), which led to uplift and the building of new areas of land.

Each cycle led directly to the next cycle.

Discussion point

Gould highlights that it was actually Lyell's rewrite of geological history that required a "hero" who used to work in the field to build his theories, to portray the "emergence of scientific geology as the victory of Uniformitarianism over the previous fruitless speculations based on untestable catastrophes". Lyell chose Hutton to be his hero.

The normal view of Hutton – as portrayed in most geology books – is that he used field observations as a basis from which he developed his ideas. He is therefore upheld as the first empiricist in geology, who devised his theory based on observation and reason and is therefore often regarded as the founder of modern geology. As Geikie noted:

> In the whole of Hutton's doctrine, he vigorously guarded himself against the admission of any principle which could not be founded on observation. He made no assumptions. Every step in his deductions was based upon fact, and facts were so arranged as to yield naturally and inevitably the conclusions which he drew from them.

Not everyone viewed him as a field-based geologist; Cuvier classed him as something of an "armchair scientist".

Hutton presented his essentially complete theory to the Royal Society of Edinburgh on the 7 March and 4 April 1785. The first full version was published in 1788 and the two-volume version appeared in 1795.

However, as we saw in Chapter 4, Hutton did not see his first unconformity at Loch Ranza until 1787 and his second one at Jedburgh later the same year. He visited the unconformity at Siccar Point, which is probably the most famous of the three, in 1788. When he presented his theory in 1785, he had previously only seen one poor exposure of a granite contact and it was not until later that year that he saw one of his key locations. It is clear therefore that he developed his theory before he had seen an unconformity or a good example of a granite intrusion.

Hutton actually used his observations to confirm rather than generate his theory. He never misrepresented his intent. He viewed the Earth as a body with a purpose. He thought that the Earth had been constructed as a stable platform for life, particularly for humans, which in itself imposed certain requirements on his theory. He wanted to find a way of ordering the Earth's complex history as a "stately cycle of repeating events" and hence required it to have a long history. He was insistent that the Earth's history led nowhere – in other words, each cycle was essentially the same as any other. He did not use strata or fossils as a sign of a history; remember that nowadays fossils are the key to a progressive history. At the time, things such as mass extinctions were not really understood.

Discussion point

Gould notes that although Hutton described all the features of unconformities, including referring to older and younger strata that show evidence of a progression, he did not talk about them in terms of their differences being tied to particular ages. He quotes Hutton as writing:

> We are not at present to enter into any discussion with regard to what are primary and secondary mountains of the Earth, we are not to consider what is the first, and what the last, in those things which now are seen.

Most textbooks imply that it was due to the "difficult" language Hutton used, that his ideas were generally disregarded, but as Rudwick writes, "Hutton was no rejected or persecuted genius," as many of his contemporaries held similar views; the main issue was the way in which he "combined them in an unusual and original way". Many savants (the word savant is derived from the French for knowing and is used for a wise or expert person) at the time held similar ideas. These types of views were widely discussed throughout the "scientific" community. Hutton's ideas were therefore not unusual and were treated with the "respect" they deserved. Rudwick continues:

> While most of them found it highly implausible, they rejected it on grounds quite other than those commonly supposed … Hutton's work has been misunderstood because it has not been treated, as it was by his contemporaries, as yet another "system" within the well-established genre of geotheormism.

The most surprising thing about the modern treatment or interpretation of Hutton's work is not his contemporarys' reaction to the notion of a vast time scale but his implication that all Primary and Secondary (stratified sedimentary rocks) had been, in Rudwick's words, "more or less completely melted or fused while buried on the ocean floor". This implies that all marine sediments were involved in at least some degree of metamorphism.

Although Hutton thought that the Earth's surface was gradually being worn down by erosion, with the resulting sediments transported and deposited in the sea over a long period of time, he was convinced that it would not be possible to observe such slow processes in action. As Rudwick puts it, "far from inferring a vast time scale from observation, Hutton deduced it from first principles and then explained away the awkward fact that its effects were unobservable".

Discussion point

How does this fit with the usual description of Hutton's work?

Hutton thought that Buffon's ideas, that the Earth had gradually cooled from a molten mass (Chapter 1) and that during the process of cooling it had cracked, forming the mountains, were not "founded on any regular system, but upon an irregularity of Nature, or an accident supposed to have happened to the sun".

Two of Hutton's key statements that set out the need for an extended time period are "the natural operations of the Earth, continued in a sufficient space of time would be adequate to [produce] the effects which we observe" and "it is necessary, in the system of the world, that these wasting operations of the land should be extremely slow". Hence, he thought that there was "no vestige of a beginning" because of the continuous recycling of the original material, and that there was also "no prospect of an end", because the operation of the natural laws could not terminate themselves or the planet.

Discussion point

Gould observed that Playfair "modernized" Hutton's views by "toning down his hostility to history". He also made his description more picturesque and dramatic.

Gould also notes that:

> Hutton followed the tradition of ordering the sciences by status from the hard and more experimental (physics and chemistry) to the soft and more descriptive (natural history and systematics). Geology resides in the middle of this false continuum and has often tried to win prestige by aping the procedures of sciences with higher status, and ignoring its own distinctive data of history. This problem, born of low esteem, continues to our day.

Simon Winchester describes this in his book on the San Francisco earthquake of 1906, entitled *A Crack in the Edge of the World*, as "old geology" born primarily in the 18th century, which, unlike its:

> … sister sciences – chemistry, physics, medicine and astronomy – never truly left the era of its making. Since the beginnings geology was a field mired in some alluvial quagmire, defined by dusty cases of fossils, barely comprehensible diagrams of crystals and the different kinds of breaks that were made in the Earth's surface (as well as by unlovely teutonic words like graben, gabbro and greywacke), and explained with cracked-varnish wall roller charts showing how the world may or may not have looked at the time of the Permian Period.

He adds that "the New Geology" is based on the new technology and thinking of "the science of the space age", which has provided us with new ways of studying the Earth.

An alternative view is presented in Tony Hallam's *Catastrophes and Lesser Calamities: The Causes of Mass Extinctions.* He writes that:

> While geology, like other sciences, has benefited enormously from technological advances in learning about the structure and distribution of rocks by using remote sensing methods, much can still be learned and needs to be learned, using time-honoured methods. The geologist's traditional tool of the trade, the hammer, remains as important as ever to those of us who are not always working in laboratories or peering at computer screens.

Hallam also records that when asked about what constitutes science, Rutherford replied that "there was physics, chemistry, which is a sort of physics, and stamp collecting". This meant that biology and geology were therefore "airily dismissed".

Think about the following questions:

Is this strictly true, are the "old ways" of observation and recording data obsolete, or do they still form a valuable part of geology today?
How do you think not being "up there with the other sciences" affects how we "do" geology today?
Do these types of views still affect the way in which geology is viewed as a science?
Is there still a credibility crisis with geology's apparent status as a "soft science"?

Geology is still fortunate in being different to most of the other sciences, in that it still allows everyone – no matter who they are, professional, amateur, or even curious day-tripper – the chance to discover something new (Chapter 9).

Charles Lyell (1797–1875) (Fig. 5.12), who was born the year Hutton died, was described by William Buckland, who taught him at the University of Oxford, as "England's first great academic geologist".

Lyell is usually portrayed as a hero of geology, someone who wrestled geology from the domination of armchair and theologically tainted speculation and made it into a modern science based on sound reasoning and observation in the field. He is presented as a person who collected field data to support Hutton's doctrine; "the present is the key to the past", a term that was actually coined by Archibald Geikie. It is said that Lyell refused to accept the existence of any process that could not be shown to be operational at the time. He also thought that the rates of geological processes had not changed significantly over time.

It has been said that his *Principles of Geology*, published between 1830 and 1833, was not a textbook but more of a "legal argument, or a piece of

Fig. 5.12 **Charles Lyell**

advocacy", that summarized all prevailing knowledge; it essentially presented a single argument. Lyell originally studied law and worked as a lawyer for two years before turning to geology; he was therefore able to use his skills in verbal persuasion to build the image of himself as the great founder of geology. In his book, *When Life Nearly Died: The Greatest Mass Extinction of all Time*, Michael Benton reports that "Sedgwick in his Presidential Address to the Geological Society of London in 1831 said that Lyell uses the language of advocate".

Lyell proposed his own view of what would later become known as uniformitarianism, by reviving Hutton's ideas of a steady state Earth, so that he could use it to challenge the views of the catastrophists. He would not entertain any ideas that rates of change or the relative importance of geological agents could have been different to those that could be observed today.

He attacked those who believed in any cooling Earth theories as these appeared to uphold the same views as the catastrophists, i.e. that geological history had been more violent in the past and was therefore different to the present. He argued that all past theories were based on a difference between past and present processes and felt that, rather like darkness versus light or good versus evil, the old ideas held back the progress of geology. He also thought that empirical observations of the Earth would allow geologists to overcome the superstitions that meant that the past had to be different.

Lyell believed that Earth movements were continually elevating and depressing the Earth's surface and were therefore gradually creating and destroying continents. If more of the Earth's surface was exposed, it would gradually warm up, and if more of it were covered with water, it would gradually cool down. In other words, there was a never-ending cycle of climatic fluctuations. He also rejected the idea that the Earth's climate could have been significantly warmer or cooler in the past.

During his travels with Murchison in Italy, he noted the fact that the pillars of the Temple of Serapis near Naples showed clear evidence that the land had sunk and risen by as much as 10 m over a period of 2,000 years. He also noted that the lower part of the columns showed that although some of the movement was rapid, this had not disturbed the columns. He thought this proved that such movements did not have to be catastrophic. He also thought the same could be argued for the formation of mountains or could explain why shell beds can be found on the tops of mountains: they could both be explained by slow gradual changes. Lyell also visited Mount Etna where he found fossils which "were apparently recent" beneath the volcano. By determining the rate at which the volcano grew through regular eruptions, he decided that it had to be "several hundred thousand years old".

Lyell believed that geology could never rise to the rank of an exact science (see the comments above) if it allowed speculation. He thought that what was needed was "laborious inquiries", i.e. the recording of detailed observations. He felt that the catastrophists had prevented the progress of geology from becoming a proper science, as biblical time scales "could not allow the present observable, slow processes to produce the Earth we have today".

He united statements about "proper scientific procedures" with a set of his own beliefs on how the Earth worked. In effect, he said that if you believe in one you must believe in the other. He stated that geological truth must be unravelled by the strict adherence to a methodology, and established "uniformitarianism" as the "officially" accepted basis for studying the Earth.

Lyell based his theory on the following:

1. *Uniformity of Law.* This means that natural laws are constant in space and time.
2. *Uniformity of Process.* This means that processes now in operation could explain past phenomena; this is also known as Actualism. This, in reality, is the same as the so-called Principle of Simplicity, which says that you do not have to invent extra, complicated, or unknown causes to explain observable phenomena.
3. *Uniformity of Rate.* Also known as Gradualism. This states that everything undergoes a slow, steady, and gradual change.
4. *Uniformity of State or Non Progressionism.* This implies that changes are evenly distributed through space and time. This means that the Earth has always looked and behaved as it does today. In other words, the land and sea can change position but they always stay in

about the same proportions. It also means that floods, volcanoes, and earthquakes have occurred at the same frequency throughout time, and that there were no early periods of faster or greater activity.

5. He also held the view that there was Uniformity of State to life. He argued that progress in stratigraphy and therefore in life was an illusion. This implied that mammals lived during the earliest Palaeozoic times and that the only reason they were thought not to exist was simply because their fossil remains had not been found yet. He thought that life could not vary, as it was tied to the physical environment.

In his book, *Global Geomorphology*, Michael Summerfield makes the following comment:

> these multiple meanings of uniformity led to much confusion in the vigorous debate which Lyell provoked after 1830 because it was possible to accept some of the propositions embodied in uniformitarianism while at the same time rejecting others.

He goes on to say,

> in fact they had no argument with the uniformity of law and uniformity of process but, on the basis of their interpretation of the available field evidence, they firmly rejected Lyell's ideas on the uniformity of rate and uniformity of state.

He explains that the argument centred on the extent and intensity to which landscape-forming processes may vary over time and concludes that the debate has still not been fully resolved, particularly with the increasing acceptance today of the role of rare, large-scale events (see below).

It can be seen that both Lyell and the catastrophists agreed on the method of inquiry, but differed on other points. Lyell disagreed with Hutton on his ideas – as portrayed by Playfair – of a succession of processes, which meant that things could change over time. To discredit the catastrophists, Lyell highlighted the old catastrophic ideas (which by then had already largely been discredited even by most catastrophists), and tied them to quite reasonable ones, such as a cooling Earth, in order to discredit all catastrophic views.

Directionalism – the idea that processes could change over time – he felt, was unscientific and any appearance of direction (change) was due to a bias in preservation rather than real changes. He thought that older rocks were more distorted because they have received more "attention" from constant forces over a longer time rather than any changes in the intensity of deformation. Lyell also thought the progressive development of organic life was wholly inconclusive and had no foundation in fact. Species "turnover" was constant through time and maintained an unchanging complexity and diversity. In his *Principles of Geology*, he said that he was sure "huge iguanodon might reappear in the woods, and the ichthyosaurs in the sea, while pterodactyl might flit again through umbrageous groves of tree ferns". It is interesting to note, with regard to these views, that Lyell had quite a big influence on Darwin during his voyages on the *Beagle*, and afterwards when he was writing his theory of evolution (Chapters 6 and 7).

In contrast, it appears that Hutton had a sequential view of uplift, which he thought may be global and catastrophic – but there was no historical distinctness, i.e. particular processes or events only happen at particular times. Lyell, on the other hand, thought that the cycles operated locally and simultaneously, giving the Earth a timeless steadiness.

Not only did Lyell completely reject any process that could not be shown to be active nowadays, but he also rejected any ideas that the relative importance of geological processes worked at different rates than those observable today (so invoking the idea of gradualism). He therefore tied himself into a straightjacket that could not fit with apparent features of the rock record. He also excluded both temporal and local "crises", which Hutton did not. It could be argued that he was an extreme gradualist, as he only allowed for processes to operate at present-day rates and intensities.

Discussion point

In a bizarre way, some of Lyell's ideas have much in common with modern creationist and creation-science thinking – both are straightjacketed by dogma, which means that they have to fit observational facts into a doctrine or reject them.

Nowadays, it is recognized that sediments show evidence of deposition by catastrophic as well as gradual processes. This is a direct result of the work undertaken by people studying surface processes: they recognize that processes depend on the interrelationship between magnitude and process, i.e. small events may occur several times a year, while larger events may only occur once in a hundred years, a thousand years, or over longer time periods. Whilst the small events may have little influence on the geological record, occasional major events (storms, floods, earthquakes, volcanic eruptions, etc.) may have a significant impact. This means that the geological record will probably preserve a large number of high magnitude (large) catastrophic events, rather than extensive records of small-scale events that happened over long periods of time.

Recent studies have resulted in the idea of gradualism being largely rejected, but Actualism – the unity of process – is still considered valid. This is based on the premise that "no powers are to be employed that are not natural to the globe, no action is to be admitted except those which we understand and can observe". In other words, the idea of divine intervention cannot be used to explain events of the past.

Gould points out that "Lyell's gradualism has acted as a set of blinders, channelling hypotheses in one direction among a wide range of plausible alternatives". He adds that "again and again in the history of geology after Lyell, we note reasonable hypotheses of catastrophic change, rejected out of hand by a false logic that brands them unscientific in principle", and concludes "our modern understanding of geology … Is an even mixture of uniformitarianism and catastrophism".

Further reading

Is it good science to judge past events in terms of a short-term human time scale? It might be worth reading "It's the only present we've got," a chapter in Derek Ager's book,* *The New Catastrophism: The Importance of the Rare Event in Geological History.*

In the first half of the 18th century, most of the best geologists, including Sedgwick and Murchison, were catastrophists and "Lyell had to forswear his beliefs like Galileo to get his chair of Geology at that holiest of London Colleges – King's. Later, ladies were forbidden to attend his lectures because of his shocking views and he resigned". This gives us some idea that Lyell's views were considered "outrageous" at the time and were not necessarily accepted by his contemporaries. The following, to some extent, confirms this.

Sedgwick criticized Lyell's ideas at the Geological Society of London in 1831 on a number of points:

1. If the Earth had originally been molten, as the presence of crystalline rocks suggest, then it must have gradually cooled down.
2. He felt that volcanic activity could not have operated at today's intensity throughout geological history.
3. He agreed that although each of the natural laws do not change, he thought that through their numerous and complex interactions, they could lead to equally complex variations and intensity of processes.
4. He also rejected Lyell's arguments that life was absolutely cyclic.

Michael Benton's book regarding the Permian extinction, *When Life Nearly Died*, makes some interesting comments about Lyell and the battle and pressures between uniformitarianism and catastrophism:

* Derek Ager, *The New Catastrophism*, 1993, published by Cambridge University Press.

> On reflection, though, these simple stories are usually more complex. Can it really ever be the case that one group of philosophers or scientists is completely right, and the other completely deluded? Clearly not. What had happened, it now seems, was that Lyell and his supporters rewrote the history of geology. In the best political tradition, geologists became divided sharply into two camps, and labels were applied. The good guys were the uniformitarians, and the baddies were the catastrophists.

As we have seen, most of the geologists at the time could be regarded as catastrophists. However, because of his renowned persuasiveness and influence, Lyell put such people in a position where they had either to accept his ideas or be labelled as a catastrophist. It is interesting to note that the catastrophists had been correct about the concept of extinctions and changes to life on Earth, something that Lyell's view of uniformitarianism effectively precluded for almost 150 years.

In her book, *Snowball Earth,* Gabrielle Walker makes an additional observation on the progress of new ideas when she says:

> when a new big idea hits the scene, there's almost always a pattern of polarization. Though a few researchers keep a genuinely open mind, others immediately entrench into either pros or cons. These vehement souls will fight, criticize, and try to pull one another down!

She notes that often it is the quality of the "rhetoric" as much as the "robustness of the data" that promotes a particular theory, and that "science works at its best when somebody puts forward a theory and everyone else tries to pull it down". The scientific process means that a theory cannot be proved but can only be disproved and, therefore, the "longer it survives the attacks against it, the more confidence you can place in it – while never knowing for certain if it is right".

Discussion point

Ager says that we "have allowed ourselves to be brainwashed into avoiding any interpretation of the past that involves extreme and what might be termed 'catastrophic' processes".

Again, from Ager's books, *The Nature of the Stratigraphic Record* and, *The New Catastrophism* – which gives some idea of how the views of continuous, slow sedimentation of the uniformitarianist approach, are being questioned nowadays – he says that:

> I am often irritated by people talking about "continuous sedimentation". Such continuity usually only exists in the minds of sedimentologists who do not bother with palaeontological detail ... It usually means, for example, just a little bit of Ordovician followed by a little bit of Silurian, followed by a little bit of Devonian with no thought of the gaps in between.

Background

The following are just a few examples of a strict interpretation of the geological record using the concept of gradualism:

Attempts to calculate rates of sedimentation in apparently continuous deposits usually do not work, i.e. calcareous ooze deposited on the floor of the Indian Ocean indicates a rate of deposition of 0.25–10 mm per 1,000 years. There is a maximum of 10,000 m of chalk that was deposited over a period of 30 million years, leading to a deposition rate of 0.3 mm per year. It would therefore take something like 2–300 years to bury an ammonite.

Carbonates are being deposited on the Great Bahamas Bank at a rate of 0.5 m per 1,000 years, but the 6,000 m that exists indicates a rate of 4–50 mm per 1,000 years, i.e. 1/10 of the rate indicated.

The average sedimentation rate on continental shelves is 10 mm per 1,000 years. The highest known rate of deposition, found in the Gulf of Mexico, is as high as 10 mm per 100 years. By taking the thickest sequence of every rock unit throughout the world and adding them together, you end up with a total thickness of rock (since the start of the Phanerozoic) of about 60 km. This would give an average rate of deposition of approximately 300 mm per year, which is obviously well in excess of the highest rate given above.

The sedimentary sequences we see today really only represent a fraction of the amount of sediment deposited and subsequently removed before it could be preserved. Any correlation chart gives the impression of a long sedimentation record with only a few gaps, but in reality what we have is a record of large gaps with the occasional sedimentation preserved; to quote Ager, it is "a lot of holes tied together with sediments". In other words, the sedimentary record we see is no more than a partial or fragmented record of what was deposited throughout geological time. Ager also puts it in another way when he compares the stratigraphic record with the life of a soldier: "long periods of boredom interrupted by moments of terror".

How much time does a break in a sequence of rocks represent? Figure 5.11 presents photographs of a number of breaks or "gaps" in rock sequences; these include the unconformity at Assynt (a), which represents a time gap of around 950 million years, and Siccar Point (b), which is 800 million years. The break at the boundary between the Carboniferous and Permian strata on Arran (c) represents a mere 40 million years, whilst the K-T boundary exposed at Stevn's Klint, Denmark (d and e) represents a time gap of probably only 100,000 years. Interestingly, the K-T boundary within the chalk deposits of the North Sea comprises a conglomerate (f) rather than the famous "fish clays".

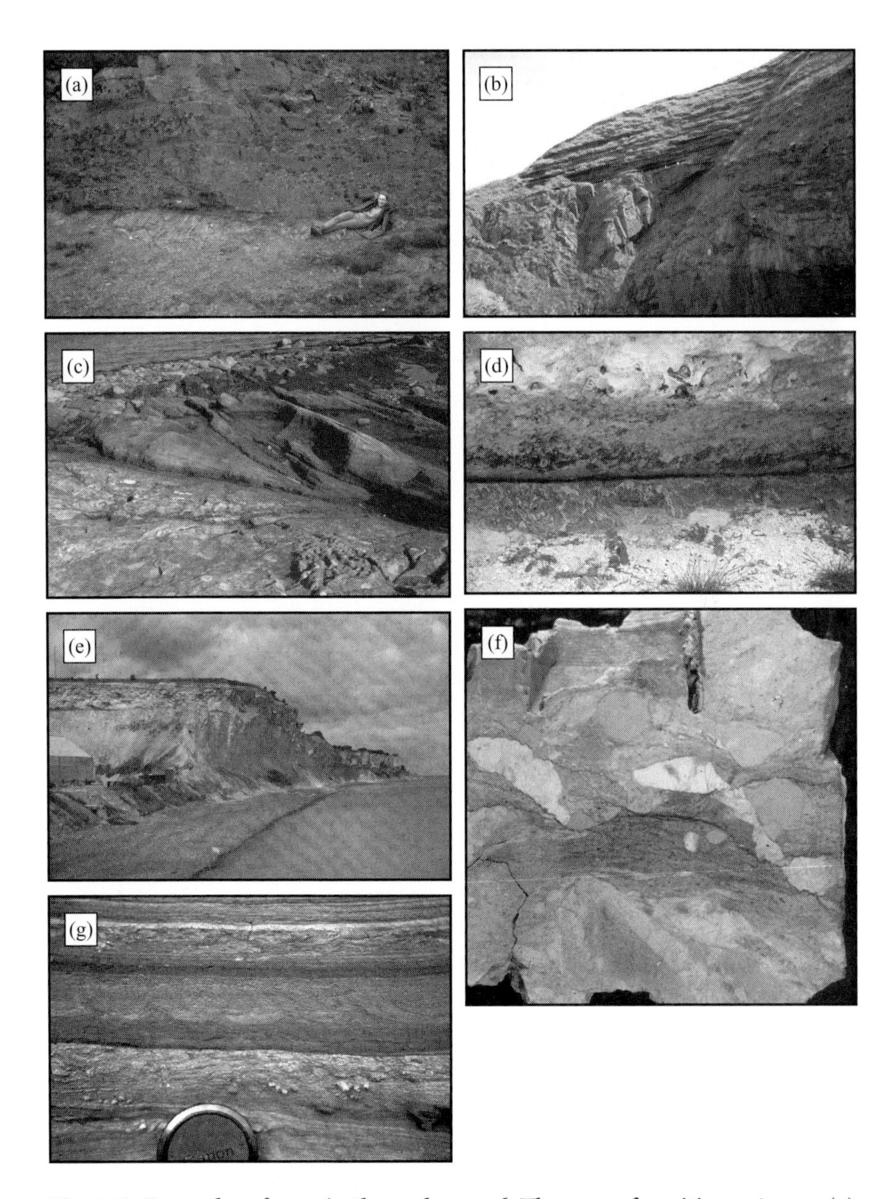

Fig. 5.11 **Examples of gaps in the rock record. The unconformities at Assynt (a) and Siccar Point (b). The boundaries between the Carboniferous and Permian strata on Arran (c) and the K-T boundary exposed at Stevn's Klint, Denmark (d and e), and in chalk cores from the North Sea (f). A sequence of sands silts and clays exposed at Scarborough, North Yorkshire (g)**

You only have to visit the coast over a period of time to see this process in action. A beach here on the northeast coast varies in height by as much as 2–3 m over a period of three to four years, depending on prevailing conditions. Each change in height represents the build-up and loss of a significant volume of sediment, of which only a fraction may be preserved in the geological record.

Discussion point

Looking at Fig. 5.11(g), how much time is represented by each of the bedding planes in these sediments at Scarborough?

The modern view of uniformitarianism recognizes that large-scale, or unusual events occur today, and have done so in the past with different intensities and frequencies. In their book *Geology: An Introduction to Physical Geology*, Stanley Chernicoff and Donna Whitney ask the question: "Have the geologic processes we see operating today occurred in exactly the same way throughout geologic time?" In answer to their question, they write, "This principle is useful for interpreting geological history and, in many cases, it's probably correct". Similar views can be found in many recent geology and Earth Science textbooks, for example, in Edward Tarbuck and Frederick Lutgens' book, *Earth: An Introduction to Physical Geology*, they comment that:

> Today the basic tenets of uniformitarianism are just as viable as in Hutton's day. Indeed, we realize more strongly than ever that the present gives us insight into the past and that the physical, chemical and biological laws that govern geological processes remain unchanging throughout time. However, we also understand that the doctrine should not be taken too literally.

Whilst James Monroe and Reed Wincander's *The Changing Earth: Exploring Geology and Evolution* tells us that uniformitarianism does not require the rates and intensities of geological processes to be constant through time, it should however be remembered that, even though these physical processes might vary in intensity and frequency, "the physical and chemical laws of Nature have remained the same and cannot be violated".

It is clear that nowadays uniformitarianism does not exclude short-term, infrequent, large-scale events such as meteorite impacts, volcanic eruptions, earthquakes, landslides, tsunamis, hurricanes, or flooding. Some geologists view Earth history much like the idea of punctuated evolution (Chapter 6), comprising short-term, large-scale events interspersed with longer periods of slow steady-state change or even relative stasis, which is similar to Ager's view (see above) with regard to the stratigraphic record (and the life of a soldier). However, even with these changing views, it should be remembered that during its early life, the Earth was subject to some processes and events that do not occur today. We know that during the early Precambrian, there were no rivers, oceans, or even continents and the planet had a poorly-developed atmosphere that was significantly different to that of today.

Discussion point

It is important to think about the relevance or usefulness of average sedimentation rates (or any other averaged process), and what they actually represent.

What is a significant gap in deposition/time, when every bedding plane is really a break in sedimentation?

As there is a move towards including an ever increasing number of "catastrophic events" into uniformitarianism, what does this mean to us today?

If Lyell or Hutton were alive today, how do you think they would react to these new interpretations of uniformitarianism?

The following section contains brief notes on some aspects of "modern catastrophic" ideas and are dealt with in a similar manner to those in Chapter 2. If you would like find out more, you should delve into recent books and journals, which cover the areas of sedimentology, stratigraphy, and geomorphology. There has been a tendency in recent publications in these three subject areas to show that natural systems tend to preserve the unusual, rare, large-scale, or "formative" events.

Discussion point

What would you consider "normal" in geology and how large or exceptional does an event have to be, to be classed as a catastrophe?

Would your answers to the above be different if you considered them over different lengths of time?

It is probably worth watching the news to see what is classified as a catastrophe and how often they occur nowadays. From this, you can start to think about and look for evidence of them in the past.

In chapter 3 of his book, Walter Alvarez* looks at various aspects of the "battle" between a uniformitarianist and catastrophist views of geology today. In the following paragraphs, you will see how breakthroughs in one area of geology were, to some extent, disregarded, whilst equally important breakthroughs in other areas supported the *status quo*. It is worth reading the entire chapter, titled "Gradualist versus Catastrophist".

Alvarez opens his discussion with a now typical comment:

> Geology could not become a real science until the stranglehold of biblical chronology was broken. Geologists have long attributed this breakthrough to two scientific heroes. The first of these was the 18th-century Scotsman, James Hutton, who is credited with the discovery that the Earth is enormously ancient. The other was the 19th-century Englishman, Charles Lyell, recognized as the father of "uniformitarianism" – the view that all changes in Earth history have been gradual. Although these traditional accounts are now recognized as oversimplified and misleading, they were accepted until recently by most geologists and palaeontologists.

He rightly points out that the key to understanding the rocks, landscape, processes, etc. depended on "generations of geologists" measuring, describing, and mapping rocks over the whole Earth, in order that we may have some idea of their distribution.

Why should this task be so important? He highlights the fact that detailed geological mapping, over a long period of time, has increased our understanding of processes such as the formation of the mountain chains by large-scale horizontal movements along thrust faults rather than large-scale vertical movements (Chapter 8). Mapping has also led to the discovery of most of the world's oil and mineral reserves. He rightly concludes that much of the last and present century's achievements are founded on geological resources discovered by geological mapping. However, in the long term Alvarez points out that, although historically, geology appeared to be lagging behind other sciences, the years of detailed mapping has given us comprehensive knowledge of significantly large parts of the Earth's land surface and ocean floor that enable the subject to "emerge as a mature science" fit for the challenges ahead. This seems to be rather encouraging compared to Winchester's views that were included above; however, Alvarez includes a sting in the tail when he adds:

> As students of geology, learning the skills of field mapping, we absorbed the traditional, exclusive focus on slow, gradual processes. We were proud that our discipline had made one fundamental contribution to the edifice of science … the principle of uniformitarianism.

Alvarez's particular interest is the Cretaceous/Tertiary (K-T) Boundary. He points out that with Lyell's view of uniformitarianism, the rate of biological change remains the same, and he was forced to conclude that an enormous period of time had to have passed across the K-T boundary in which no rocks are preserved, to account for the huge change in life forms on either side. Interestingly, according to Alvarez, Lyell thought that the K-T boundary represented a longer period of time than from then to the present (now considered to be 65 million years).

As we have seen, uniformitarianism therefore became fundamental to geological thinking and anyone with a catastrophic view could face major opposition. Alvarez includes a good example of the pressure this change exerted when, in the 1920s, J. Halen Bretz of the University of Chicago, described a network of huge, dry channels in the eastern area of the State of Washington. He thought that these channels looked like large-scale river valleys and proposed that catastrophic flooding by glacial meltwater had formed them. His views were considered too reminiscent of a biblical flood and uniformitarianists went to great lengths to dismiss his ideas. As far as they were concerned, any catastrophic ideas were deemed unscientific rubbish. Alvarez describes how "uniformitarianism dogma" continued to discount Bretz ideas for over 20 years, until similar features were discovered on Mars. Fortunately, the story had a happy ending when, in 1965, following a field visit to the area, an international group of geologists proved that Bretz, now aged 83, had been right. It is said that they then sent him a telegram in which they said, "we are now all catastrophists".

Even in the late 1960s, when the Plate Tectonic revolution began, geologists believed that positions of the continents were fixed (Chapter 8). The introduction of this new theory affected almost all aspects of geology. It made use of the years of "routine" detailed mapping (as described above) to draw together information collected from different continents, which showed that Plate Tectonics was a gradual, uniformitarianist theory. This reinforced the idea that catastrophic events could still be largely disregarded as old-fashioned and could be ignored, thus maintaining the *status quo*. Nowadays, geologists are increasingly identifying or re-interpreting many geological sequences as more unusual, rare events and in doing so are eroding the traditional uniformitarianist view of geology.

Further reading

It is interesting to note from the above how rigidly following a uniformitarianist approach delayed understanding some aspects of geology. It is well worth reading the whole of Alvarez's *T. rex and the Crater of Doom*, as it shows over and over again how Nature throws in red herrings, which led scientists in the wrong direction because they had a particular idea that hindered them from seeing any alternatives from the evidence they already had to hand.

Your Inner Fish by Neil Shubin also provides a good example of the planning and the role of serendipity in scientific discoveries.

Discussion point

The problem is how do you study anything with a completely open mind? Most people have a purpose or reason for studying anything that is usually based on prior knowledge or understanding. (See catastrophes and the nature of science later on in this chapter, and the development of Plate Tectonics theory in Chapter 8.)

As we saw in the introduction to this chapter, Alvarez asks questions such as, what kind of a past has it been? Is Earth history a chronicle of upheavals, catastrophes, and violence? Alternatively, has our planet seen only a stately procession of quiet, gradual changes?

It can be seen that geology is slowly moving from a strict uniformitarianist approach to a mixture of gradualism and catastrophism, where both concepts form the opposite ends of a spectrum of Earth processes. Geologists continue to find that most changes in Earth history have taken place slowly and gradually but occasionally it has suffered enormous catastrophes, which have totally redirected the subsequent course of events.

Discussion point

Given enough time, a rare event becomes a probability, and given a long period of time, it becomes a certainty. For example, there have been 200 large tsunamis in the last 2,000 years – this implies that there could be as many as 100,000 every million years. It has been calculated, given the frequency at which they occur, that a hurricane will pass over any particular point in the Gulf of Mexico once every 3,000 years. To us – when we see them on the television or read about them in newspapers, tsunamis and hurricanes are extreme events, but over an extended period of time, could we class them as normal, regular events?

Does normality therefore depend on a particular perspective of time?

5.5 Mass extinctions

It is not the intention to discuss the causes of mass extinction here or to go into any detail, as they are covered in many geological textbooks. There are also a number of general science books that cover either particular extinctions or the subject area. Three of the more interesting are Michael Benton's *When Life Nearly Died*, Walter Alvarez's book, *T. rex and the Crater of Doom*, and Michael Boulter's book, *Extinction: Evolution and End of Man*. The first two are particularly good, as they take the reader through the background to extinctions in general. They also cover two of the most important extinctions, namely the Permian extinction (Benton), the largest in history, and the Cretaceous-Tertiary extinction that killed off the dinosaurs (Alvarez). A fourth book, *Catastrophes and Lesser Calamities: the Causes of Mass Extinctions*, by Tony Hallam, provides an extremely good overview of all of the extinction events in which he presents some of the latest ideas and evidence for their causes and effects.

Before we look at the nature of extinctions, it is worth noting the following comments by Palmer:

> Neither Murchison nor any other mid-19th-century geologists spotted any sign that the end of Permian times was marked by the biggest extinction event in the history of life … It is not as if Murchison and his contemporaries were not concerned about the changes in fossils from one group of strata to the next; they were. It was because it was the presence of such changes that they were using to justify distinct "systems" of strata.

In other words, they had identified major changes in the fossil record but were not particularly concerned by the causes or processes involved in those changes. As we have seen from the previous section, gradualism had largely replaced catastrophic ideas, and therefore the concept of extinction did not sit comfortably with slow gradual changes. In other words, extinctions were dismissed as an out-of-date idea linked with "old-fashioned, backward-looking" views of the progression of life. Like so many other things in science (as Alvarez shows clearly in his book), if you are not looking for something, you tend to either miss or dismiss evidence for it.

It is estimated that over 99% of all known life has become extinct in the past. There are two types of extinction: those in which one species evolves into another, so that the original effectively disappears; this is referred to as a pseudo-extinction; the other (which is the one we would normally think about when using the term extinction) is where a species dies out completely. On occasions, something happens to cause a large number of species to disappear, and these are known as mass extinctions.

The five major mass extinctions, which are normally recorded in textbooks, occurred at the following points in geological time:

1. Late Ordovician
2. Late Devonian
3. Late Permian
4. Late Triassic
5. Late Cretaceous.

These are often referred to as the "big five", two of which, the Late Permian and the Late Cretaceous, marked huge changes in life on Earth, so much so that the former marks the end of the time period named the Palaeozoic Era (meaning "old" or "ancient" life) and is the largest of the five, and the latter marks the end of the Mesozoic Era ("middle" or "intermediate" life).

There is also evidence that there have been 19 periodic extinctions with what appears to be a periodicity of about 26 million years. Could these have been the result of similar events to that proposed by Alvarez for the Cretaceous-Tertiary (K-T) extinction, namely an astronomical event? Benton notes that responses to this proposal "were polarized. Many enthusiastic geologists and astronomers accepted the idea, and its clear implication that a regular periodicity in mass extinctions implied a regular astronomically controlled causative mechanism". Whereas others argued that "each extinction was a one off, and that there was no linking principle". He then goes on to say that the current view is that:

> Most palaeontologists and geologists have just quietly let it drop. Close analysis of the fossil data has failed to confirm periodicity. Indeed, scrutiny of some of the extinction peaks, such as the three in the Jurassic, has suggested that these are largely artefacts of the data collecting.

It is interesting to see that Boulter asks an important question – could all mass extinctions be explained by a single cause? He notes that some geologists had already suggested that the history of life, including phases of diversification and extinction, might be controlled by local or global changes in temperature or in sea level. In other words, external controls play an important part in the progress of life. Geologists generally view life as just a part of the Earth system, whereas biologists tend to consider that changes are driven by internal controls. As we shall see, this difference in view is also relevant when discussing evolution and the progression of life in Chapters 6 and 7. To some extent this is due to a difference in the perception of the speed of change; geologists view changes over hundreds of thousands or millions of years as fast; whereas fast to a biologist may be decades, hundreds, or thousands of years.

Of all of the extinction events, interest in the K-T extinction has always been high, due to the loss of the dinosaurs but – as Tony Hallam points out – "unfortunately the fossil record for the dinosaurs is so patchy and limited that it is difficult at present to say much of note about such event". He goes on to explain that even though concrete evidence is limited, it has not stopped numerous scientists (not just geologists) from proposing their own hypotheses for the demise of the dinosaurs. He explains that Alan Charig, a former curator of fossil reptiles at the Natural History Museum in London, has found more than 90 hypotheses, most of which were "more or less fanciful". These included:

1. Climate change
2. Disease
3. Nutritional problems
4. Various parasites
5. Infighting
6. Hormonal changes
7. Slipped discs
8. Racial senility
9. Dinosaur eggs being eaten by mammals
10. Changes in the sex ratios of embryos, due to variation in atmospheric temperature
11. Suicidal psychoses
12. Inherent problems of their small brains.

Hallam also includes two other hypotheses:

> Perhaps the most fanciful of all appeared in the late 1980s in a letter to the *Daily Telegraph*, from a scientist respectable enough in his own field. He thought that the dinosaurs had died out as a consequence of an AIDS infection induced by viruses introduced from outer space.

A further favourite relates to the decline of the naked seed plants, or gymnosperms, at the expense of the flowering plants, the angiosperms at the end of the Cretaceous Period, which implied that, the "herbivore dinosaurs died of constipation". Unfortunately, as Hallam points out, the main change in flora took place about 35 million years before the Cretaceous/Tertiary boundary. He adds that the extinction at the end of the Cretaceous Period appears to be the culmination of numerous small-scale extinctions across a wide range of organisms rather than a sudden, dramatic collapse.

I can remember being told in the 1980s by Jake Hancock that the dinosaurs, rather than being wiped out at the end of the Cretaceous, were "on their last legs, most having gradually died out before then". New and future fossil finds may well reveal similar patterns within the Late Ordovician, Late Devonian, and end-Triassic extinctions.

Boulter makes an interesting comment on how scientists work and how they usually present topics such as extinctions to the public:

> The whole scientific way of thinking is based on the challenge of proving something wrong, on refusing to accept the conclusions of others and hopefully being able to prove the hunch right. Journalists and teachers do a bad job in conveying this conflict to members of the public, let alone to politicians. Both groups want straight answers to straightforward questions and don't understand it when they can't get them.

Discussion point

This is one of the reasons for writing *Time Matters* – to have the opportunity to show how complicated and often convoluted the path of science can be, and that frequently scientists do not have the definitive information the public require. They often fail to recognize that the pubic are not interested in the nitty-gritty details of academic arguments, and fail to present key points in a way that most people can understand.

Extinctions appear to play an important role in biodiversity. It has been said that life on Earth needs extinctions for it to change and diversify and therefore, high observed extinction rates may simply reflect a high evolutionary turnover of species. Extinctions are also often only identified in rock sequences with an exceptional degree of fossil preservation. For example, the fossil record indicates that a series of extinctions occurred between the start of the Cambrian and the Early Ordovician.

In most cases, the interval of floral and faunal diversity loss does not correspond to the interval of maximum extinction. Many extinctions span a significant period of time, for example, the Permian extinctions spanned 10 million years. However, this time span may be, to some extent, an artefact of the large marine transgression that occurred at the end of the Permian. The same problem may also account for the Late Triassic extinctions. The Late Devonian extinctions actually cover a broad period of approximately 25 million years, from the Middle to Late Devonian. The Wenlock and Ludlow Series in the middle of the Silurian Period contain a broad peak of small-scale extinctions that lasted for approximately 15 million years, similar to those found in the Lower Carboniferous.

There is also clear evidence that some extinctions occurred over a relatively short period, examples of these including Late Ordovician and Late Cretaceous events.

Finally, in addition to the "big five", there is increasing evidence that a Late Precambrian extinction (650 million years ago) may have killed approximately 70% of then existing life.

As well as the 26-million-year cycle of extinction, others have suggested that extinctions have occurred every 32 million years over at least the last 250 million years (based on evidence from ammonites and foraminifera). This regularity has been attributed to extraterrestrial driving forces such as comet collisions, an undetected tenth planet that has a highly irregular orbit, or even the existence of an "undetected binary solar companion" (a death star), which has a highly elliptical orbit 0.3–3 light years from the Sun.

There is, however, a general lack of correspondence between predicted and observed extinctions. This may be resolved as more fossils are unearthed, particularly from areas of the globe where, historically, there has been little geological investigation.

Discussion point

In his book *Earth Time; Exploring the Deep Past from Victorian England to the Grand Canyon*, Douglas Palmer writes:

> the whole problem of such extinction events is still intriguing and important for our understanding of the history and evolution of life. Do not believe all you read about such events. Even experts, who have devoted the best part of their working lives to the study of major extinctions, are still trying to assess the details and put together workable scenarios.

Having looked at uniformitarianism and catastrophism, should we regard extinctions, no matter what the scale of their severity, the time period over which they occurred, or their apparent regularity, as catastrophic or uniform processes?

5.6 Alternating warm and cold conditions

This is not just a case of changes associated with the last glaciations, or changes in today's or tomorrow's climate due to global warming – it refers to changes that have occurred throughout a significant part of geological history.

The rock and fossil record clearly show that during at least the last 600 million years the Earth's climate has alternated between periods of generally warm and cold conditions. These have been termed Greenhouse and Icehouse, where overall global temperatures resulted in the presence of either non-permanent or permanent polar ice caps. Under Icehouse conditions, the presence of ice has also been identified at latitudes as low as 45°. The mechanisms that determine which climatic state reigns are, at present, poorly understood. A number of factors have been suggested, including the global positions of the Earth's continents, the existence and position of supercontinents, differing levels of atmospheric CO_2, variations in solar radiation, and changes in relative sea level. It is likely that the change from Icehouse to Greenhouse or Greenhouse to Icehouse conditions, may well be a combination of a number of factors, some of which we do not yet have the data to identify.

Whatever the cause, the Earth's climate has alternated six times between these two stable conditions during Late Proterozoic and Phanerozoic times:

1. Greenhouse conditions from Early Cambrian to Late Ordovician, which peaked at 468 million years ago.
2. Icehouse conditions again from the Late Ordovician to Early Silurian, when ice sheets covered North Africa, which at the time was over the South Pole.
3. Greenhouse conditions returned between the Early Silurian to the Early Carboniferous, 428 to 333 million years ago.
4. Icehouse conditions affected the Earth once more from the Early Carboniferous to the Late Permian (333 to 258 million years ago), with the greatest extent of ice between the periods 315 to 296 million years.
5. The most recent Greenhouse conditions existed from the Late Permian to the Tertiary, 258 to 55 million years ago.
6. We are now in Icehouse conditions that have included the episodes of the last ice age.

The above shows that, overall, the Earth's climate, at least for the last 600 million years, has been predominantly (~80%) under Greenhouse conditions, and that therefore the present Icehouse conditions are not the global norm.

Further reading

In chapter 10, *The Stratigraphic Record and Global Rhythm*, of their book *The Key to Earth History; An Introduction to Stratigraphy*, Doyle, Bennett, and Baxter provide a very good overview of geological climate change (from which the above has been taken), possible control factors, and its effect on the evolution of life.

You should be aware of the current debate about global climate change and the suggestion that humans are in the process of bringing about a major change in the climate. If the geological record provides evidence of previous changes in climate, not least of which are the changes between the glacial and interglacials of the Pleistocene over the last 1.5 million years, how significant to the Earth's long-term future is our effect?

The International Panel on Climate Change (IPCC) reviewed the history of global climate changes over the last 900,000 years and showed that average global temperatures varied between 5 °C below and 2 °C above our present climate. If we go back further than 3 million years ago – apart from a period covering the later half of the Carboniferous Period and the first half of the Permian Period – since the start of the Cambrian Period, global temperatures have always been between 2 and 7 °C higher than today.

Discussion point

Does this help put the current debate on climate change into geological perspective?

If you look at the world's climate on a geological time scale, we are in a generally cold period; in Icehouse conditions. Therefore, when we compare current changes in the climate to historical data to predict the future, we generally do so on the basis of accurate records, which are less than 300 years old. Interestingly, a significant proportion of these were collected towards the end of the Little Ice Age, a generally colder period in recent climatic history, which only officially began in about 1310, and ended in around 1850. Is it therefore surprising that temperatures are rising?

It is important to state quite clearly here that I am not saying that we are not affecting the climate – I think that we are. You should treat the origins, influences, and content of any discussion of the effects of climate change with due caution and consideration. It is clear from many examples in this and other books that science is not necessarily transparent and that different individuals or groups have different reasons for conducting their scientific research. Do not forget that there is big money riding on each of the views, as well as some big reputations at stake – some of which will inevitably be proved wrong. Climate change is already a big business, which will only get bigger in the future. (See Chapter 9 for a further discussion.)

5.7 Catastrophes and the nature of science

Having looked at the different views of uniformitarianism and catastrophism, and the relatively recent acceptance that catastrophic events, including extinctions, form an important part of the geological record, it is worth thinking about how science and scientists deal with challenges and changes to their often long-held ideas.

In his book *Science on Trial: the Case for Evolution*, Douglas Futuyma* makes the following comments about scientists, how they think, and how they work:

> Science is science only if it limits itself to determining the nature of reality. The hallmark of science is not the question "Do I wish to believe this?" but the question "What is the evidence?". It is this demand for evidence, this habit of cultivated skepticism that is most characteristic of the scientific way of thought. It is not limited to science, but it isn't universal either.
>
> … At its best, science challenges not only non-scientific views but established scientific views as well. This in fact, is the wellspring of progress in science. Our knowledge can progress only if we can find errors and learn from them. Thus, much of the history of science consists of a rejection or modification of views that were once widely held. Geologists once believed in the fixity of continents, but now believe in continental drift. The Newtonian theory of physics is now seen as a special case of a larger theory that includes relativity. Scientists realize, if they have any sense at all, that all their currently accepted beliefs are provisional. They are, at present, the best available explanations, but subsequent research may show them to be false or incomplete. I cannot stress this point too strongly. Unlike fundamentalists who will not consider the possibility that they could be wrong, good scientists *never* [his italics] say they have found absolute "truth". Read any scientific paper and you will find the conclusions couched in words like "apparently" or "it appears that".
>
> Scientists accept uncertainty as a fact of life. Some people are uncomfortable unless they have positive, eternal answers; scientists come to terms with uncertainty and mutability as a fundamental condition of human knowledge. Science is not the acquisition of truth; it is the quest for truth.
>
> The picture I have just painted is, of course, a somewhat idealistic one. In fact, scientists are just as human as anyone else. They believe that one or another hypothesis is most likely to be true, and they engage in sometimes bitter battles to defend their ideas. Scientists' beliefs are also shaped by their political, social and religious environment.
>
> … Thus the common image of scientists as abstracted, unbiased, detached intellectuals has no foundation in reality. Scientists are often highly opinionated, even in the face of contrary evidence; and they are often not particularly intelligent either. The spectrum of scientists, as of any other group of people, runs from the brilliant to the fairly stupid. Almost every scientist has made more than one asinine statement in the course of his or her career, and some make them habitually.
>
> If scientists can be just as biased, subjective, and foolish as anyone else, why should we have any belief in what they say about physics, evolution or the causes of cancer? Because scientists are motivated not only by a

* Douglas Futuyma, *Science on Trial: The Case for Evolution*, 1995, Sinauer Inc.

> quest for knowledge but a quest for reputation. And there is no better way for a scientist to achieve reputation than to demolish existing ideas by finding contrary evidence, or to propose a theory that explains the evidence better. This means that although individual scientists often make errors, the body of scientists in a field eventually uncover these errors and attempt to correct them. Every scientist's research depends on the research of others in the field; so out of pure self-interest, every scientist scrutinizes the work of others carefully, to be sure it is reliable. Science is a self-correcting process.

Snoke in his book, *A Biblical Case for an Old Earth*, tells the story of the scientists who claimed to have discovered "cold fusion". He relates that:

> ... eventually the cold fusion claims were disproven to the satisfaction of most scientists (a small group of scientists continue to pursue this effect to this day). In the aftermath, the great majority of scientists felt the original scientists had engaged in unethical, or "pathological", science. What made it pathological was not that they were wrong; many scientists get things wrong all the time. But, these scientists bypassed the normal scientific avenues of fact checking and went straight to the public with their claims.

Discussion point

Science tends to act as a self-regulating system, whereby someone proposes an idea and other people test it. Obviously, the reasons for doing so may be flawed, as we can see from the examples above (as well as many more that have been included throughout this book), but in the long run the checks and balances tend to make sure that the "truth", or at least the best attempt at the "truth", at any given time wins through.

Has this changed your views of scientists and the way they work?

Snoke – a physicist, university professor, Presbyterian elder, and an Intelligent Design (ID) proponent – uses a similar argument with regard to the claim by young Earth and "creation-scientists", that scientists are involved in a conspiracy over the age of the Earth. He notes that some people hold the belief that a significant proportion of non-Christian scientists are part of a large-scale conspiracy to fabricate geological data, thus making it compliant with their view of an ancient Earth. This disregards the fact that a great many geologists are involved in exploration to locate natural resources. Without the accuracy of their records and careful interpretation of geological data using well-founded models, such resources would not be found. It is only because their models and theories fit the data that they can make use of them. They are not concerned with "creating a vast religious deception", even though "some Christians fault the old-earthers for violating the scientific method because they deal with things that lie in the past, and therefore beyond the realm of falsifiable predictions". Snoke makes it clear that this is completely untrue as, for example, Plate Tectonics theory, an old Earth concept, proves to be a very successful, predictive theory. He explains that:

> … just as capitalism tends to make people work towards productive goals out of self-interest, so it also tends to keep them scientifically honest, since a person who consistently denies realities and makes false predictions of where to drill for oil, at a cost of millions of dollars, will not last long in the business.

In her book, *The First Fossil Hunters: Palaeontology in Greek and Roman Times*, Adrienne Mayor recounts the words of Thomas Kuhn in his book *The Structure of Scientific Revolutions*, that science works on the basis of maintaining the *status quo* or at least steady-state change interspersed with "conceptual revolutions" that replace older views. He adds that:

> … when scientists are engaged in gathering data on normal phenomena, the resistance to noticing discrepancies is strong … "Normal science does not aim at novelties of fact or theory, and when successful, finds none." If anomalies begin to be perceived in great enough numbers, however, they cast doubt on expected patterns, and anxiety arises over the failure to make them conform to the accepted paradigm. Then, when sufficient anomalies accumulate to induce repeated crises within the scientific community, the scientists struggle to adjust, and revolutionary scientific advances can occur.

In other words, generally speaking, scientists are conservative in their approach, and resistant to change.

Douglas Palmer, the author of, *Fossil Revolution: the Finds that Changed our View of the Past*, makes the following observation based on the problems encountered in trying to get the Precambrian Ediacara fossils – originally found in Canada in 1946 – into the scientific arena:

> making an important discovery is not enough to guarantee its appreciation, unless it is made by someone who is already famous and has publishing access to the most important scientific journals. Publication does not necessarily help if it is in an obscure journal and the scientist is not part of an appreciative network of scientists who will promote the find.

As William Ryan and Walter Pitman put it in their book, *Noah's Flood: The New Scientific Discovery about the Event that Changed History*, "the test of a good scientific theory is its power of prediction".

Although this may be true, in her book *Flat Earth: The History of an Infamous Idea*, Christine Garwood observes that nowadays people are far more suspicious of the motives behind science and scientists, noting that from the 1960s, the public mood changed. They were no longer necessarily prepared to view scientists as objective and working for their good. She adds that:

> In America especially there was a heightened public criticism of the role of experts: radicals on the New Left had gone so far as to portray scientists and engineers as the secret rulers of the industrialized Western world … their hold on society seemed unbreakable: their knowledge was so sophisticated that it was impenetrable to the majority, and science had become compartmentalized to such a degree that even experts in related fields could not grasp the nuances of their colleagues' work.

Garwood suggests that the basic question we need to ask, is does society possess adequate knowledge to trust without doubt what someone else, who is portrayed as an expert, is telling us? Alternatively, do we, and we frequently have to, take them at their word? It is interesting to note that we are going through a period, particularly relating to our litigation culture, in which experts, and expert witnesses are coming under increased scrutiny and their opinions are being increasingly challenged.

Discussion point

You may ask, why has society's questioning of science and scientists been included, why is it important? It questions the basis on which scientists study and view their own subject, and the whole of science in general. If they come under attack, they have to be able to justify themselves, their views, and the way they reach those views, to be able to avoid falling into the "traps" of others. Garwood's observations also shows how the public's attitudes towards science and scientists have changed over time; something that scientists should always be aware of. If the public do not understand what scientists are trying to say and do not trust them, why should they believe them?

The answer to the above question is that if they wish the public to trust them, there is increased pressure on scientists to be as transparent in their work and provide as clear and understandable an explanation as possible.

Further reading

It is worth reading chapter 2 of Snoke's book, in which he reviews the scientific basis of his argument for an old Earth.

Having looked at how science and scientists operate, let us go back to the uniformitarianist versus catastrophist debate and the idea that the present is the key to the past.

As mentioned above, in his book *The New Catastrophism*, Tony Ager includes an interesting chapter entitled, "It's the only present we've got". Here he looks at how reliable our "present" is as a guide to the past. Ager says that:

> The basic principle of uniformitarianism is, of course, that we can use the processes going on at the present time to interpret the events of the geological past. We must ask ourselves, however, whether our ephemeral present is typical of the infinite number of fleeting presents that have passed in the course of Earth history. It may be that it is an odd and atypical present that we have to use to try to understand the past … A great deal depends on what we mean by the "present". When did our "present" start? Is the key provided by that present long enough to unlock all the difficult doors of the past?

As Ager says, when Lyell wrote his *Principles of Geology*, and laid down his definition of uniformitarianism, the Pleistocene was unknown and very little was known about the Lower Palaeozoic and Precambrian. Ager raises the question, "one wonders if a much earlier Chinese Lyell or an Inca Lyell would have thought the same way". He notes that:

> Many early humans must have seen geological phenomena far more violent and spectacular than we know in historic times, including the last great volcanicity across northern Europe from the Auvergne to Romania and the explosion of Santorini which may have given rise to the Atlantis legend. In New Zealand, the first Polynesian immigrants may have seen and suffered some of the last huge volcanic explosions in North Island. Looked at the other way around, we must ask ourselves if our present is really all that typical and we must always accept the basic constraint that it may be a very odd period in which we now live.

With regard to conditions during the early Precambrian, Ager recalls that:

> … with ideas such as a less bright sun, a faster rotation of the Earth (leading to shorter days of 15 hours), smaller landmasses and a greater proportion of CO_2 in the atmosphere (and hence a strong greenhouse effect) the Earth was very different during Achaean times.

In the more recent geological past, he highlights the fact that:

> The organic world has played an important part in changing the processes which affect the surface of our planet. Thus it must have been a very different place before the evolution of grass produced a close ground cover, thus reducing the rate of surface erosion and providing the food for many groups of newly evolving mammals.

Commenting on the effects of glaciations Ager adds:

> the most obvious way in which our present is atypical of most of the record is that we live in the aftermath of a great series of glaciations. Glacial theory with huge ice sheets covering much of the northern hemisphere would have been an anathema to Charles Lyell, since it was so "catastrophic" in concept, though he was well aware of the importance of changes in climate and recognized the probable existence of glaciers where there are none today. In fact most of the leading geologists of the day, such as Murchison, rejected the idea.
>
> … there are several features of our present world which clearly relate to the later stages of a major glaciation. One is the generally low sea level due to water being retained in the ice caps. So we have greater lengths of rocky shorelines today which are rarely seen in the geological record.

Even the relative shortage of carbonate deposits forming at present is probably a result of the recent glaciation, "since carbonates seem to go with global warming". Continuing with the climate change theme, Ager comments that "in our short-sighted way, we tend to think of our present global warming as a one-off phenomenon, brought about by man's own foolishness. However, the geological record shows that it has happened many times in the past". He concludes that:

> Man has changed his world and made it less suitable for interpreting the past, though I always argue that we are just one more species "doing its thing" like every other species before us. Every species today, and presumably in the past, has pursued a policy of "my species right or wrong". The survival of the species is more important than anything else. We are just more efficient at changing things and at killing other organisms, including our own kind.

As Ager identifies, the introduction of life produced a profound change in the Earth's climate by adding oxygen – "the first atmospheric pollutant" – and the invasion of the land by grass (see above) changed the nature of the landscape forever. There are numerous other examples of "evolutionary changes" found in the geological record, which have brought about profound changes that all go to make up the world as we know it today but that have not existed throughout geological time.

Further reading

Brian Fagan's book, *The Long Summer: How Climate Changes Civilisation*, makes interesting reading with regard to how we have been affected by and have affected the Earth's climate.

Discussion point

Having read the previous sections, it is worth thinking about the following questions:

Is uniformitarianism as a concept still valid in geology today?
How did the views of Hutton and Lyell differ with regard to uniformitarianism?
What are the implications of these differences in today's view of geology?
How has hindsight influenced our view of 18th and 19th century catastrophists?
What is the validity of catastrophism today?
How reliable are present processes as keys to the past?
Why does the stratigraphic record contain more gaps than rocks, and what effect does this have on our views and interpretations of Earth history?

Nowadays, Plate Tectonics provides a unifying theory with regard to most aspects of modern geology. As all three were "forward-looking" geologists, how do you think knowledge of this theory would have influenced the views of Hutton, Werner, and Buckland?

5.8 Palaeogeography and Earth history

Before moving on to the next chapter, it is important to think about one of the ways in which all of the areas of geology that we have looked at so far have been drawn together – namely through palaeogeography. We have seen how geologists determine the age, relative and absolute, of rocks, and how remnant magnetism in rocks has been used to reconstruct the movement of landmasses and ocean floors. We also looked at how the study of stratigraphy, and geological sequences enabled early, pre-evolution theory, geologists to determine the divisions of geological time. We have seen the process by which extinctions were identified and their importance understood, and how large-scale alternations of the global climate have been identified. All of these provide a consistent story of change over an extremely long period of time.

We come back to the famous phrase "the present is the key to the past", even though we recognize rare and unusual events. Here in Britain, there is a consistency in the patterns of processes and depositional features that cannot be denied. There have been periods when we were covered in deep oceans, enormous deserts, tropical rain forests, shallow tropical seas, and vegetation-covered planes. We have been caught up in the formation of vast mountain ranges and the opening up of the Atlantic Ocean and North Sea. Each of these events is clearly and demonstrably represented and preserved in the rock record.

Using this information, it is possible to produce stunning, detailed maps, such as those contained in the *Atlas of Palaeogeography and Lithofacies* published by the Geological Society (Memoir No. 13). Here the work of 35 contributors has resulted in the production of 102 beautiful, detailed maps with supporting summaries that show the palaeogeography of the UK over the last 1,050 million years. During this period, the UK and its original components have travelled the face of the Earth through different climatic zones because of Plate Tectonics (Chapter 8). Similar information can be found in books, such as the Geological *History of Britain and Ireland* edited by Nigel Woodcock and Rob Strachan, and the first edition of *The Key to Earth History* by Peter Doyle, Matthew Bennett, and Alistair Baxter. In their second edition, Doyle and Bennett exchange maps of the UK for maps that cover Europe. Similar maps covering North America have been included in such books as *The Changing Earth: Exploring Geology and Evolution*, written by James Monroe and Reed Wincander, or Kent Condie and Robert Sloan's *Origin and Evolution of Earth: Principles of Historical Geology.*

Why are these so important? They clearly show that all sedimentary rocks are not laid down in a haphazard manner. Their internal structures and spatial distribution demonstrate that processes and conditions operating today deposited them. People such as those working in the oil industry are able to determine the location, depth, and spatial distribution of oil and gas reservoirs using this type of information – an activity that would be extremely difficult, if not nearly impossible, if sediments throughout much of geological history had been deposited under vastly different conditions.

Discussion point

To end this chapter, here are two interesting views on geology, geologists, and catastrophism:

In his book, *The New Catastrophism*, Derek Ager adds this "disclaimer":

> In view of the misuse that my words have been put in the past, I wish to say that nothing in this book should be taken out of context and thought in any way to support the views of the "creationists" (who I refuse to call "scientific").

I hold similar views about *Time Matters*.

The second view comes from the book *The Genesis Flood* (a well-known Creationist book by Whitcomb and Morris (1961), which makes interesting reading). "Instructed Christians know that the evidence for full divine inspiration of the scriptures are far weightier than the evidence for any fact of science". They feel that it is impossible to harmonize uniformitarianism with the Flood and therefore either the biblical record of the Flood is not true or the way in which geological history is interpreted is wrong and has to be changed. They go on to say a total re-evaluation of historical geology should be undertaken in order to harmonize it with the biblical record and conclude with:

> It will likely have to be attempted, if at all, by men outside the camp of Professional Geologists. It is unlikely that many students majoring in the field could survive several years of intense indoctrination in uniformitarian interpretation of geology without becoming immune to any other interpretation and still less likely that they would ever be granted graduate degrees in the field without subscribing wholeheartedly to it.

Any scientist, not just geologists, should look at all the facts available and try to interpret them as best we can, hopefully without any preconceived ideas or constraints. Our views may change as new evidence is found and we should therefore never be afraid to question existing ideas. Comments like those of Whitcomb and Morris should only emphasize that geologists and other scientists should take care to consider every aspect of their subject carefully and thoroughly in order that such accusations cannot be justifiably levelled against them.

In his book *The New Creation: Building Scientific Theories on a Biblical Foundation*, Paul Garner, reviews a number of the geological topics covered in *Time Matters*. He includes a number of new creationist-based theories to explain these phenomena. These include an explanation for a post Noah's Flood single Ice Age (chapter 15) that lasted perhaps 500 years rather than the generally accepted multiple events over a period of 2.6 million years. It is worth thinking about the implications of this for any discussion of man-induced climate change.

6
Evolution

6.1 Introduction

This chapter can be used as an introduction to the historical development of the theory of evolution and the changes that it has been through in recent years. As with Chapters 4 and 5, this chapter and Chapter 7 go hand-in-hand. It is particularly important, when you go in search of fossils, to think about what they are, what they represent, and what they tell us about life in the past. Whenever anyone finds fossils it is always interesting to remember that it is probably the first time in millions or possibly hundreds of millions of years that anything has seen that particular animal.

Further reading

If you would like to read an interesting book which covers the debate over evolution, I would recommend *Science on Trial: The Case for Evolution* by Douglas Futuyma. Chapter 4 is particularly worth reading, but there are a wide variety of other books on the subject, because it is still something of a contentious matter with a variety of different groups of people.

Although evolution is primarily a biological theory, its implication's have a significant impact on geology and palaeontology (the study of fossils). Equally, geology has had a large impact on the theory of evolution, as fossils provide much of the evidence that life has changed through Earth history. As we have seen, geology also provided the time scale and time frame in which evolution has been generated.

Time Matters: Geology's Legacy to Scientific Thought, 1st edition. By Michael Leddra.
Published 2010 by Blackwell Publishing Ltd.

It should be noted that the possibility of preservation of the remains of a particular organism is dependent on a specific set of conditions. This generally means that the chances of an organism being preserved as a fossil are an exception and not the rule. The conditions that existed in ancient seas, lakes, rivers, and particularly terrestrial areas in which organisms lived, rarely allow their remains to be preserved, due to decomposition, alteration, erosion, abrasion, chemical dissolution, or predation.

This means that the fossil record has to be biased towards the preservation of particular organisms that have a better chance, through their make-up, lifestyle, and habitat, of being preserved. It also means that when we look at any particular section of the fossil record, we are only being given a glimpse of the diversity of life that probably existed at the time. Having said that, every so often palaeontologists find a sequence of rocks in which the conditions were good for preserving a wider range of fossils than they would normally expect to find. When this happens, the diversity of life revealed is almost inevitably wider than expected. These are called *Lagerstatten*, which roughly translated means "fossil bonanzas". A table (Table 7.1), which contains the sites of most of the worlds known *Lagerstatten*, has been included at the end of Chapter 7.

Discussion point

What effect do you think the diversity of fossils found at these sites would have on our view of the fossil record and how could this affect our view of life in the past?

In her book, *The First Fossil Hunters: Paleolontology in Greek and Roman Times*, Adrienne Mayor relates ancient myths to the abundance of large fossil bones found around the Mediterranean and adjacent areas. She also shows how various natural philosophers at the time were constrained by their own ideas and traditions of how animals formed and changed (Chapter 7), and makes the following point:

> What survives of philosophical writings strongly suggests that, for whatever reason, the philosophers opted out of the "unknowable" problems of giant bones. But inquiry proceeded without them, resulting in natural knowledge based in experience and expressed in geomyths. The myths were not a formal theory in the modern sense, of course, but as palaeontologist Niles Eldridge observes, neither is the theory of evolution a "fact". Like the mythical paradigm, our own modern paradigm is "an idea – a picture" that allows us to explain observed facts.

The idea that living things could evolve (change) had been around for a significant time before Darwin; for instance, as Alan Cutter relates, Robert Hooke had thought about the idea that because ancient species of animals were different from modern ones they could have evolved from one to the other. Hooke saw the diversity of animals as evidence of their adaptability and Cutter quotes him as saying, "we see what variety of Species, variety of Soils and Climate, and other Circumstantial Accidents do produce". As we have already seen, Hooke also considered the possibility that gaps left by extinctions could be filled by the evolution of others species.

It was Carolus "Carl" Linnaeus (1707–1778) (Fig. 6.1) – a Swedish botanist, physicist, and zoologist known as the father of modern taxonomy – who also gave us the term *Homo Sapiens* in 1758. As part of his *Systema Natuarae*, he devised a hierarchy for all living things. He was unconvinced by the story of the Creation and Noah's Flood for the development of life and proposed, in 1774, that life had originated on a mountainous tropical island surrounded by a primeval ocean. The height of the mountain produced a sequence of climates that changed from tropical close to sea level, to polar conditions at its top. This allowed a wide variety of life forms to develop, which spread out from the island as the "Flood" waters dropped.

Fig. 6.1 **Carolus "Carl" Linnaeus**

As new organisms were discovered, his hierarchical system, which was based on shared physical characteristics, continued to expand into a system that became known as Linnaean taxonomy – which is still used today.

Georges Cuvier, considered to be the founder of modern palaeontology, is famous for his work on vertebrate fossils that led him to recognize the sequential nature of the fossil record, which he published in 1811. He studied fossils of the Paris Basin, used the idea of "lost species" to account for the features that differentiated between the different vertebrates he found, and developed the theory of "revolutions on the surface of the globe", which is of course a catastrophist approach. This meant that successive catastrophes repeatedly extinguished life, and were followed by a deluge that allowed creation to "take up where it left off", with additional improvements.

With increasing knowledge, the connections between fossils and living organisms became clearer. Eventually, to account for so many changes, Cuvier was forced to invoke an increasing number of floods until he arrived at a situation of continuous change. His followers included Alcide D'Orbigny (1802–1857), considered the father of biostratigraphy, who proposed 28 different creations in which similar fauna were created with only minor changes.

Cuvier was able to study fossil elephant bones that had been brought back to Paris from a collection in Holland. He compared their characteristics with the bones of living elephants and decided that they belonged to an extinct type of elephant, and published the results in *On the Species of Living and Fossil Elephants* in 1796. He also noted that the bones were not found in the tropical areas in which elephants now live. He therefore decided that this meant that the mammoth had no living relatives and had therefore become extinct. He named this new species the "mammoth". Shortly after this, a mammoth was discovered in Siberia preserved in the permafrost: this specimen proved that Cuvier's interpretation had been correct.

Cuvier showed that the younger deposits contained more fossils of species known today, but that as he went down through older strata, fewer and fewer "modern" species existed. In 1812, he showed conclusively that many fossils found in the Paris Basin had no known modern counterparts, thus representing extinct species.

From his anatomical studies, Cuvier proposed a four-fold "body plan" system:

1. *Vertebrata,* which included all those with backbones;
2. *Articulata,* including arthropods, insects, and segmented worms;
3. *Radiata,* including echinoderms (sea urchins);

4. *Mollusca,* which covered all invertebrates that had a bilateral symmetry.

He believed that similarities between organisms were due to common functions and not common ancestors. This implied that it was function (use) that determined form (shape), rather than form determining function. Cuvier was opposed to the ideas of Lamarck and Buffon, who both thought that morphology could be affected by environmental conditions, and did not believe in the idea of evolution as he thought that any change in an organism's anatomy would make it incapable of surviving. He did however agree that the fossils showed a gradual advance in the complexity of life. It was Cuvier who noted that the age of reptiles predated the age of mammals and that within the latter, marine mammals predated terrestrial mammals.

In his *Theory of the Earth* published in 1813, he said that he thought that the Earth was immensely old but that generally conditions had been similar to today with periodic, sudden, worldwide "revolutions", which led to extinctions. He thought that, following extinction, other organisms were created to fill the gap that had been left.

Cuvier is famous for his boast that "from a single bone, or even a portion of a bone, the anatomist can reconstruct an entire animal". Although he was opposed to the evolutionist ideas of Lamarck, Cadbury reports that Cuvier:

> ... believed that fundamental laws must govern the anatomy of creatures as surely as the laws established by Newton now govern physics. If a creature was a carnivore, Cuvier observed, all of its organs would be designed for this purpose.

His principle of "correlation of parts" meant that every element of an animal's anatomy had to be interdependent for it to be able to survive.

Discussion point

Is Cuvier's boast valid?

Jean Baptiste Pierre Antoine de Monet, Chevalier de Lamarck (1744–1829) (Fig. 6.2), was well-known for his ideas of "transformism", a term that he himself never used. Lamarck developed his theory based on the inheritance of acquired traits. Both Darwin and Lyell considered him to be a great zoologist and the forerunner of evolution theory.

Fig. 6.2 **Jean Baptiste Lamarck**

Initially Lamarck joined the army, as did his father and a number of his brothers. Following an accident, he left the service and eventually went on to study medicine and botany. He became an assistant botanist at the Jardin des Plants (the Royal Botanical gardens), which later became the National Museum of Natural History in Paris, where he was made Professor of the Natural History, of insects and worms (he is considered to be the founder of invertebrate palaeontology). It was Lamarck who introduced the term *invertebrates* to cover a wide range of animals without backbones, which had largely been neglected or ignored.

After an extensive study of fossils from relatively recent rocks in the Paris Basin he felt that, even though they varied, they showed "continuity". Lamarck thought that life showed "graduated differences", whereby all the major animal groups could be fitted into a step ladder series that extended from the simplest life forms up to mammals and then to humans. He thought that all organisms were linked together in a continuous chain which, as Rudwick says, meant that species "were in the long run no more than arbitrary points on a continuum" in which "the difference between fossils and living species might simply reflect this process of endless flux".

It should be noted that Lamarck was looking at fossils that were very similar to the present-day organisms, and had a large number of living relatives, whilst Cuvier's fossils have no modern equivalents. Both scientists were thinking along the lines of transformation of species, but Cuvier would only deal with features he could observe, so the idea of common ancestors did not figure in his arguments. In 1807, Cuvier stated that it was the task of the evolutionists to "show how living species could be related, by a sort of descent, to those of the first inhabitants of the earth, a knowledge of which is transmitted to us by their remains in the fossil state".

While Cuvier based his ideas on fossil evidence, Lamarck made almost no use of fossils in his explanation of his ideas in his *Philosophie Zoologique*, published in 1809. His area of study concentrated on invertebrate animals, which are notoriously poorly-represented in the fossil record, as they have few if any "hard parts" that constitute most fossil remains.

In 1801, Lamarck wrote, "time and favourable conditions are the two principle means which Nature has employed in giving existence to all her productions". He thought that animals responded physically to environmental changes in order to survive. This may result in an increase or decrease in the use of particular structures or organs within the organisms, which allowed them to be able to adapt to the changed conditions. This concept formed his First Law.

His Second Law stated that all changes were heritable, i.e. passed from one generation to the next. Together these laws meant that organisms gradually and continuously changed as they interacted with or in response to changes in their environments. Although he used many of the same examples for his theory, including natural selection that Darwin used, Lamarck's theory was largely ignored. He was even attacked for his ideas during his lifetime. For example, William Conybeare, a colleague of William Buckland, thought that Lamarck's idea of transmutation was "monstrous". In the end, he died in poverty and obscurity. Interestingly, it was one of his followers, Frederic Gerard, who coined the phase the "theory of the evolution of organized beings" in 1845.

6.2 Darwin and evolution

Charles Darwin (1809–1882) (Fig. 6.3) studied medicine and theology at Edinburgh University, during which he gained an interest in natural history. During his famous five-year voyage on the *Beagle*, he established himself as a very able geologist as well.

Whilst he was on this voyage, his friend, Reverend Professor John Stevens Henslow (1796–1861) – who had been a lecturer in botany at the University of Cambridge whilst Darwin was there – used to let selected naturalists have access to the specimens that Darwin had been sending back from his expedition. This meant that by the time he returned to Britain in 1836, Darwin was already a celebrity in scientific circles. In fact, he had collected so many specimens during the voyage, that he and Henslow had to arrange for a number of naturalists to describe and catalogue them so that the work

Fig. 6.3 **Charles Darwin**

could be completed within a reasonable time scale. These included Richard Owen (Chapter 7), who studied many of the fossil bones Darwin had collected. In 1837, Darwin moved to London to be closer to those who were studying his specimens so that he could supervise the work; it also allowed him to circulate in the "right" scientific circles that existed in the capital, which comprised people of significant influence.

In June 1858, Darwin received a letter from Alfred Russell Wallace (1823–1913) (Fig. 6.4), a naturalist working in Malaya, about an essay Wallace had written titled, *On the tendency of varieties to depart indefinitely from the original type.* Darwin immediately recognized that this was effectively his own theory that he had been working on for over 20 years, and that Wallace had independently come up with the same ideas and conclusions. Darwin is said to have commented to Lyell that "even his terms now stand as heads of my chapters", and "all my originality, whatever it may amount to, will be smashed, though my book, if it will ever have any value will not be deteriorated; as all the labour consists in the application of the theory". Darwin offered to send Wallace's manuscript to a scientific journal and added to Lyell that "I would far rather burn my whole book than that he or any other man should think I had behaved in a paltry spirit".

On 1 July 1858, the Secretary of the Linnaean Society read manuscripts by both authors, neither of whom were present, to an audience who had

Fig. 6.4 **Alfred Russell Wallace**

gone to hear a paper by George Bentham on the "fixity of species". The two papers effectively claimed the opposite of Bentham's ideas. As Ruth Moore says in *Man, Time, and Fossil: The Story of Evolution,* "when the reading ended, silence fell. There was no discussion. Some of those present realized they had lived through a historic moment". Darwin then prepared an abstract for publication, which resulted, within a year, in the publication of *On The Origin of Species.*

According to Jerry Coyne's book *Why Evolution is True*, "Darwin's theory that all life was the product of evolution, and that the evolutionary process was driven largely by natural selection, has been called the greatest idea that anyone ever had".

Like Lamarck, Darwin made little use of fossils, when he published *On The Origin of Species*; his evidence was almost exclusively based on living organisms. In his view, fossils did not show in detail the progressive and gradual evolution proposed by his theory. He emphasized the gaps in the fossil record more than the information they provided. Fossils, he argued, provided on a grand scale the progressive transformation of the living world – they did not provide all the evidence needed to understand the mechanisms involved. Fossils provided "snapshots" of this gigantic process. In fact, as Prothero points out, although Darwin did not use fossils as evidence for evolution, he spent two chapters of his book explaining that, although the fossil record was imperfect, it showed that evolution had occurred over a long period of time.

Discussion point

In *The Message of Fossils* published in 1991, Pascal Tassy says that "Every evolutionist after Darwin has always considered fossils to be tangible proof of evolution".

It is important to make a distinction between fact and interpretation. A fossil is a fact that can be held, observed, characterized, described, and recorded. Geologists understand that if fossils are found in their life position, the rocks in which they are found were formed at or some time after the time of death. (It is usually clear if a fossil has been moved from its *in situ*, life position.) The rocks in which fossils are found represent their stratigraphic position, unless they have been reworked. Each of the above is an observable fact, which generally does not change. Nevertheless, when you use this information to interpret the environment in which they lived and/or died and how they relate to fossils found above and below them, you are dealing with interpretation based on current theories and your own level of knowledge. You can therefore place them into a history but, as Tassy asks, "Which history? … Using the same fossils, two palaeontologists, each as competent as the other, can construct two different genealogical trees and tell two different stories". A fossil is a fact, which nowadays is normally indisputable, but Tassy adds that "as soon as it is interpreted, as soon as it becomes intelligible, it acquires a cargo of theory and thus becomes a scientific object, and by that very fact becomes conjectural and subject to dispute".

He notes that "the sedimentary and fossil record shows us the products of evolution, not the links that may have existed between them", which means that "what is older cannot be automatically assumed to be an ancestor of what is more recent".

Prothero writes that:

> Although scholars in 1859 may have considered Darwin's evidence from fossils weak, this is no longer true today. The fossils record is an amazing testimony to the power of evolution, with documentation of evolutionary transition that Darwin could only have dreamed about. In addition, detailed studies of the fossils have even changed our notions about how evolution works and have fuelled a lively debate in evolutionary biology about the mechanisms that drive evolution.

What were Darwin's basic propositions? The following are taken from *Ecology: Individuals, Populations and Communities* by Michael Begon, Colin Townsend, and John Harper, published in 1990:

1. The individuals that make up a population of a species are not identical: they vary – though sometimes only slightly – in size, rate of development, response to temperature, etc.
2. Some at least of this variation is heritable. In other words, the characteristics of an individual are determined to some extent by its genetic make-up. Offspring receive their genes from their parents, and offspring therefore have a tendency to share characteristics with their parents.
3. All populations have the potential to populate the whole Earth, and they would do so if each individual survived and each individual produced its maximum number of offspring. But they do not: many individuals die prior to reproduction, and most (if not all) reproduce at a less than maximum rate.
4. Different individuals leave different numbers of descendants – this means more than saying that different individuals have different numbers of offspring. It includes the chances of survival and reproduction of these offspring, and the survival and reproduction of their offspring in turn.
5. Finally, the number of descendants that an individual leaves depends – not entirely but crucially – on the interaction between the characteristics of the individual and the environment of the individual.

According to Coyne, the modern theory of evolution can be summarized by the following:

> Life on Earth evolved gradually beginning with one primitive species, perhaps a self-replicating molecule, that lived more than 3.5 billion years ago; it then branched out over time, throwing off many new and diverse species; and the mechanism for most (but not all) of evolutionary change is natural selection.
>
> … when you break that statement down, you find that it really consists of five components: evolution, gradualism, common ancestry, natural selection, and nonselective mechanisms of evolutionary change.

It is clear that some environments are more favourable to some organisms leaving descendants than others, and some individuals leave more descendants than others. This means that heritable characteristics can change with successive generations by "natural selection". In addition:

> past environments act as a filter through which combinations of characteristics have passed to be adapted (fitted) to their environment only because present environments tend to be similar to past environments. The word adaptation gives an erroneous impression of prediction, forethought or, at the very least, design. Organisms are not designed for, or adapted to, the present or the future – they are a consequence of, and therefore adapted by, their past.

We know that the same conditions apply to landscapes, which are also rarely, if ever, in equilibrium with present environmental conditions. If this is true, can evolution theory, and the concept of natural selection be used as a forward predictive tool, or just a retrospective model?

Natural selection was supposed to be based on "survival of the fittest", but what does this mean? Fitness is a relative and not an absolute term: it refers to those that "leave the greatest number of descendants relative to the number of descendants left" by others. Natural selection therefore cannot lead to the evolution of "perfect individuals" but only "favours" the fittest organisms available, which at any point in time and space could be a fairly restricted choice. In other words, it is a process that leads to the development of the fittest available and not "the best imaginable".

Where organisms live is frequently the result of an "accident of history". That is, they are there because of such things as continental movement due to Plate Tectonics, climate change, or the development of isolated island communities, etc.

Darwin provided a significant body of evidence for evolution that, as Coyne says, "convinced most scientists and many educated readers that life had indeed changed over time. This took only about 10 years after *The Origin* was published".

By the end of the 19th century however, Darwin's theory of adaptation by natural selection had fallen into the background. Both biologists and palaeontologists based their work on "anti-Darwinian" ideas, such as those of Lamarck, as they offered a method by which evolution would follow a linear trend rather than be driven by local, unpredictable events.

Discussion point

Could this be partly based on the move towards a preference for a uniformitarianist (predictable) rather than a catastrophist, i.e. random or unpredictable point of view?

Darwinism was revived in the 1920s, when the science of genetics began to develop. During this period, most people believed that evolution was a process directed by producing features that were either useless or beneficial to the organism, with no regard to its environment.

By the 1940s, the theories of non-adaptive evolution started to be discarded. The combination of genetics and Darwinism began to dominate every science, particularly with regard to the way in which animals behave in response to their environment.

Originally, the science of genetics was used to create models of evolution that bore little resemblance to the actual behavioural patterns observed in Nature.

Now evolution is generally viewed as an unpredictable sequence of events, governed by random mutations and the hazards of an ever-changing local environment. Every natural population contains a large gene pool in which changes in the environment can cause a once useless gene to be "switched on"; enabling representation of the characteristics produced by that gene to become more prevalent.

Evolution theory is based on simplicity, but this is complicated by convergence of the independent appearance of identical characteristics in different species. Convergence implies that similar evolutionary events can be repeated; it also means that different species can share a single characteristic inherited from a common ancestor carried through in a single evolutionary event. It is possible that parallel evolution can occur, but this is deemed to be less efficient and so is normally rejected.

The gradualist model implied that evolution is a gradual and slow process that advances over a long period of time. If breaks in the evolutionary sequence are found, they may be attributed to gaps in the stratigraphic (rock and fossil) record. Superposition (see Steno in Chapter 2) allows us to see the "direction" (i.e. the chronological order) of evolution, which develops as progressive morphological changes with the occasional backward trend. This means that recent members of a group of organisms are linked to their oldest relatives by a chain of intermediate forms. It is noted, however, that on a number of occasions the idea of a gradual progression in the development of life are found, on closer inspection, to be unfounded – for example, we could use Micraster, an echinoid (sea urchin) and a type fossil of the Upper Cretaceous. Originally, changes in the shape of these fossils were thought to have occurred over a long period of time. However, recent studies have shown that the changes in shape were due to the influx of species from surrounding sedimentary basins and were not due to gradual changes of the existing, local population.

This example shows that apparent evolution was not necessarily due to a gradual change in a population. For a modern example, we could use the plight of the red squirrel. In the future, people finding the fossil remains of red and grey squirrels may well decide that the apparent replacement of red squirrels by grey squirrels in the stratigraphic record indicated that the smaller red squirrels evolved into larger grey ones. In reality, we know that the grey "outsiders" are gradually taking over the habitats of the native red squirrels.

Discussion point

Could this be viewed as a case of survival of the fittest?

6.3 Punctuated equilibrium and geographic speciation

In 1972, Niles Eldridge and Stephen Jay Gould proposed the idea that evolution occurs as a result of "punctuated equilibrium", whereby Nature proceeds through a series of periods of stability interspersed by phases of rapid transformations. Because this is significantly different to the idea of slow, steady change, they have been accused of being anti-evolutionists. The basis of their argument could be put as "evolution is possible because, for most of their existence, species do not evolve". This means that when conditions remain stable there is no need for an organism to change. However, when environmental conditions alter, organisms have to change relatively quickly in order to take advantage of the changes or simply just to survive. Just as Ager described the stratigraphic record as comprising "long periods of boredom interrupted by moments of terror" (Chapter 5), a similar description could be used for Eldridge and Gould's theory.

Another variation on the theory of evolution that has been developed is that of "geographic speciation": this implied that new species develop via local, peripheral populations located around the edge of the geographical location of the parent or main species. Initially, the peripheral and main species would be the same through interbreeding, but if for some reason the peripheral species are isolated from the main group and their environmental conditions change, they would evolve to be able to adapt to those changes. If at a future date the barrier between the peripheral and main groups was removed, re-established interbreeding could introduce the adaptive changes that had developed in the peripheral species to the main species.

Discussion point

Darwin proposed a similar idea before he wrote *On The Origin of Species*. He described it as the way in which groups could be isolated and allowed to develop without coming into contact with the parent species. However, by the time he came to write his book, he felt that it was unnecessary to propose the idea of isolation as a process in evolution. He was then criticized by Moritz Wagner, a German naturalist, who insisted on the need for a period of geographic isolation.

Geographic speciation does not allow for progressive, *in situ* evolution to occur. This rationale has been used by punctuated evolutionists who say that evolution is dominated by rapid changes in the peripheral species, which feeds into the main or parent species, interspersed with longer periods of equilibrium. Eldridge and Gould stated that "these long periods of stasis are the dominant feature of species development". This concept has been supported by research into theoretical genetics under the heading, "adaptive landscapes". The punctuated equilibrium model says that speciation is a sudden event taking place over a period of between 5,000 and 50,000 years. On the geological time scale, this is rapid and would be difficult to recognize in the fossil record, but this is of course slow in terms of a biological time scale. Palaeontologists have also pointed out that gradual transformations over millions of years are also recognizable.

If periods of speciation are short, major morphological changes could occur without leaving any discernable trace in the fossil record. The fact that they also occur in marginal populations in marginal environments means that the chances of these changes being preserved and found are slim. This is not an easy way out of explaining away fossil evidence of intermediate forms.

Not all evolutionary developments, driven by genetic-environmental conditions, can play an important role in inhibiting or activating changes. If an environment remains constant for a long time, a species may become very specialized and dependent on a limited range of environmental conditions or a particular source of food. If that environment then changes, the species may not have sufficient genetic variation to adapt to the change fast enough to survive.

6.4 Intermediates – what are we looking for?

One of the major problems with finding evidence for evolution is that the fossil record rarely contains intermediate forms, i.e. ones that show a transition between anatomically distinct organisms. The fact that intermediate life forms are rarely found is often used as one of the main arguments against evolution.

During the late 19th and early 20th centuries, palaeontologists viewed evolution as a linear process and they were therefore committed to looking for the "missing links" in the fossil record. They linked fossils together into patterns that showed evolutionary trends developed over long periods of time. A good example of this is Charles Walcott (1850–1927) (Fig. 6.5), who found the Burgess Shale in Canada, which contains a huge variety of often-unique fossils. These fossils belong to the Cambrian Period when life on Earth rapidly evolved, a process that is often referred to, incorrectly, as the "Cambrian explosion". When the original stratigraphic column was being put together (Chapter 3), the rocks assigned to the Cambrian Period were thought to contain the earliest evidence of life – hence it represents the start of the Palaeozoic Era. Walcott's initial classification of the fossils was based on preconceived ideas of evolutionary trends and it was not until relatively recently that many of the fossils have been identified as unique "trial" life forms, which appear nowhere else in the fossil record. This may well be true, but it might also be, in part, an artefact of the "normal" fossil record that such life forms, which have a limited stratigraphic span, are not

Fig. 6.5 **Charles Walcott**

preserved. Having said that, if you follow that argument to an extreme you could end up with a view similar to Lyell's – in other words, that they existed throughout geological history but their fossils have not been found anywhere else yet.

Further reading

If you would like to find out more about this unusual fossil sequence, Gould's *Wonderful Life: The Burgess Shale and the Nature of History* make a very interesting read.

One of the key things that stratigraphic "snapshot" sequences, like the Burgess Shale, emphasize through their amazing level of fossil preservation is that the "normal" fossil record often only provides a very limited view of the variety of life that exists at any particular period in geological history.

It may be that intermediate fossil species may be difficult to identify, let alone find. The traditional view of "missing link" fossils is that they should be, in some way, a half-and-half organism, showing the characteristics of both sides. This implies that the transition fossils between fish and amphibians, or reptiles and birds, or even reptiles and mammals should all contain elements of both.

However, the fossil record shows that it is always one characteristic or one group of characteristics that evolves at a given rate rather than the entire morphology of an organism; particular features evolve in response to particular environmental changes, while other features remain the same. This leads to a situation where an organism will display an association of characteristics that show a variety of evolutionary stages, a process that has been termed "mosaic evolution", where fossils show a range of "primitive" and "complex" features. They do not exhibit half-and-half features.

Evolution is due to natural selection and mutation, as all populations contain extensive "gene pools" by which genetic variations can continuously arise. Animals also have the ability to respond to chemical signals by evolving features they had previously lost or that have remained dormant for a long time.

A new mutation may be no better or worse than the existing species, but random events can have a bigger impact on small populations than on larger ones. This ultimately leads to the extinction of the smaller group. Whether a mutation is "good" or "bad" depends on the environment: it is possible that a "bad" mutation could exist for long enough that changes in the environment could lead to that mutation becoming "good". Equally, some mutations are neither "good" nor "bad", as they have only a secondary effect or no effect at all on the species.

Discussion point

How realistic is this as a proposition?
Are transitional forms likely to be half-and-half animals?
Is it more useful to look for "key features" that exist in both groups?

"Missing links" can be found, but usually we are looking for the wrong features. The differences we should be looking for are based on anatomical changes, which in turn are based on specific inputs. Many evolutionary changes are the result of chance – a particular response to a particular situation. A change in, for example, the climate, ecology, or geography, may generate a different evolutionary response. The problem with this concept – known as "contingency" – is that it reduces the predictive value of evolution. As you look back at the geological record, the differences between some species become increasingly blurred. It is misleading to think that "advanced species" will be "advanced" in all of their features or attributes. Every major gap in the fossil record of related species exhibits one or more characteristics that may be termed "evolutionary novelties", which cannot be related to predicted patterns of development. Most evolutionary developments can be accounted for by simple changes in the relative development of particular features, i.e. differences in the growth rate of different parts of the body.

Discussion point

Very few questions about evolution can be answered without fossils, as they show us the evidence of change over a significant period of time. As we find more fossils, some of the gaps will be filled or models of particular animals will be refined or re-thought – even ones that appeared to have been agreed on and set for a long time. A classic example of this was the proposed linear evolutionary development of the horse. Originally, the fossils were placed in an order, which showed that as they evolved they grew in size, etc. Nevertheless, as more fossils were found, the view of their evolution changed from a purely linear process to one in which both large and small species lived together.

Remember, some groups of animals are only known through fossil teeth, small bits of bone, or even trace fossils. If additional, more complete fossils were to be discovered, our interpretation may change. Conodonts are a good example of this.

Further reading

Neil Shubin's excellent book, *Your Inner Fish*, provides many examples of intermediate fossils, features, and traits, together with a very good discussion of the role, nature, and evidence of "missing links".

Other books that provide interesting examples of "missing links" and transitional fossils include: *For the Rock Record: Geologists on Intelligent Design*, edited by Jill Schneiderman and Warren Allmon; Mark Isaak's *The Counter-Creationism Handbook*; Donald Prothero's *Evolution: What the Fossils Say and Why It Matters*; Denis Alexander's *Creation or Evolution: Do We Have to Choose*?; and Jerry Coyne's *Why Evolution is True?*

The following are a few extracts from chapter 4, "The Fossil Record", of *Science on Trial* by Douglas Futuyma, concerning dating and fossils with regard to evolutionary development:

> With this method [radiometric dating], geologists have obtained the dates for the geological ages as shown in the geological time scale ... They have dated ancient rocks, moon rocks and meteorites, and found a consistent age for the solar system ... Astronomers have postulated that because of tidal friction, the rate of the earth's rotation has slowed down at a rate of two seconds every 100,000 years, so that a Palaeozoic day should have been about twenty-one hours long. Corals lay down a layer in their skeleton every day, as well as layers that mark the passage of years. John Wells of Cornell University reasoned that Devonian corals, if they lived 380 million years ago, should have about 400 daily layers per year in the Devonian. And so they had; the estimate of the age of the Devonian, deduced from coral skeletons, corresponds perfectly with the estimate from radioactive dating.
>
> ... The other claim of the creationists is that we have no way of being sure that the rate of radioactive decay has always been constant, even if it seems to us now to be immune to any outside influences ... The very same processes of atomic changes that result in radioactive decay are those that enable us to build atomic bombs and nuclear reactors. The physics of these processes is very well understood – perhaps too well.
>
> ... There is immense regularity in the fossil record. Mammoths, dinosaurs, and trilobites aren't mixed together at random. From the beginning of the Cambrian, hundreds of millions of years pass before the amphibians appear; then another hundred million or so years until the first reptiles, and another hundred million until the first birds. Without any reference to the fossil record, taxonomists have claimed, using the principles by which they construct phylogenic trees, that modern mammals are descended from primitive shrewlike insectivores, primitive mammals from reptiles, reptiles from amphibians. These judgements come entirely

from anatomical studies of living species. We predict, then that amphibians, reptiles, primitive mammals, and modern mammals should appear in sequence in the fossil record, and they do. It is impossible, if evolution is true, that any mammal fossils should ever be found in Devonian rocks, and indeed there are no such fossils.

... The fossil record is not, of course, a book that we can open at will to look up dates and historical figures. It is the accumulation, by hard work, of fragments of early life that happen to have been preserved and happen to be found ... The fossil record is a source of endless frustration. The museums of the world hold millions of fossils, but they are from rich beds found here and there, a scattering of fragments from the vast expanse of time and space.

Moreover, we know that the recovery of any organism from the past depends on a concentration of improbable events: the organism must [generally] have hard parts that resist decay; it must be buried in sediments that happen to become solidified into rock; the rock must escape erosion and metamorphism for eons; and it must be exposed in places where geologists happen to find it ... Poor as the fossil record is, however, it tells us that there is an orderly history of life. Different groups originated at different times, not all at once ... The rocks tell us, also, that extinction is the fate of almost all species. Moreover that the rate of extinction doesn't slow down as time goes on; recently evolved species have no longer tenure on Earth than ancient ones ... Whether or not the fossil record reveals gradual evolution is very much a matter of scale.

Discussion point

Having looked at the nature of fossils and the development of a number of evolution theories, think about the following questions:

In order to explain evolution, do we need to find – and therefore should we be looking for – the missing links?

Recent theories indicate that rapid evolutionary changes occur in marginal environments; how would this affect our quest for "missing links" or transitional fossils?

How has the study of genetics helped us to understand the theories of evolution?

What is the difference between gradualism and punctuated evolution?

Concepts of evolution have changed significantly in some areas of the natural sciences since the theory was first introduced by Darwin. On what evidence are some of those changes based?

Does evolution require a long geological history?

Should there be an order to the development of life?

7
Evolution versus Creationism

7.1 Introduction

This may seem a rather odd title to a chapter in a book about geology, but as you have already seen from several of the pervious chapters, many areas of geological thought have developed from biblical-based ideas. As with these other topics, the study of the origins of life and its development through geological history has not been without its conflicts and casualties. You will also find in this chapter that some of the battles are still being played out by people standing their ground along traditional lines.

Although many of the battles between evolutionists and creationists may appear to be along the lines of historic discussion, they are probably just as important now as they were in the past. Today there is a growing push in some schools in this country – usually with a religious foundation – to teach evolution and creation in equal measures. This growing trend is also evident in America, where the re-emergence of creationist ideas has even led to the sale of at least one creation book at the Grand Canyon (Fig. 7.1).

Discussion point

Grand Canyon: A Different View, by Tom Vail, uses selective examples to "prove" that the geological sequence within the canyon and the formation of the canyon itself can be explained by Noah's Flood. He appears to deliberately ignore, for instance, all unconformities other than the "Great Unconformity" (its location is signified by the X in Fig. 7.1). He omits the presence of lava flows within the sequence and highlights missing formations within the canyon as evidence for erosion during the flood, without mentioning that they exist in the surrounding area.

Time Matters: Geology's Legacy to Scientific Thought, 1st edition. By Michael Leddra.
Published 2010 by Blackwell Publishing Ltd.

Fig. 7.1 **A view of the Grand Canyon from the south rim. The position of the "Great Unconformity" is indicated by an X**

Chapter 6 looked at the historical development of evolution, which initially made little use of fossils. This chapter begins with a review of the historical understanding of the nature and relevance of fossils to help appreciate the background for the arguments between evolution and creation.

Let us start with a quote from the introduction to Douglas Palmer's book, *Fossil Revolution: the Finds that Changed our View of the Past*:

> The discovery of fossils and their scientific meaning was a shocking business that radically altered our sense of ourselves and our relationship to other life forms. From thinking of ourselves as being one rung below the angels on the ladder to heaven, we have found that we are genetically over 98% chimp and just one of a dozen or more human-related species.

He explains that the fossil record probably represents less than 1% of life that has existed on Earth – an important concept to bear in mind throughout this chapter.

As we have seen in Chapter 6, the fossil record is usually limited to life forms that lived in particular environments and were composed of materials that made them more likely to be preserved. It is thought that "between several hundred million and a few thousand million species have exists over the last 540 million years ... of which only a few hundred thousand are known as fossil forms". This means that there could well be a significant number of new species yet to be "dug out of the rocks".

Discussion point

What do fossils represent?
What do they really tell us about the development of life?

7.2 Fossils

Early humans must have come across objects that we nowadays refer to as fossils. It is commonly thought that they probably would have had little idea as to what these objects really were, but an interesting book by Adrienne Mayor, *The First Fossil Hunters: Paleontology in Greek and Roman Times*, presents strong evidence that this is not necessarily true. In her book – which is well worth reading – she shows that in the Mediterranean, and areas as far away as Kazakhstan and China, ancient civilizations were finding and interpreting the origins of large fossil bones. In fact, there appears to have been a large-scale trade in such items from at least the 8th century BC and some trade as early as the 22nd century BC. She puts forward a very good argument to show that ancient ideas of giants and the like may well have been based on the discovery and interpretation of such bones, which in some areas were extremely common.

She reports that the world's first palaeontology museum was established by the Emperor Augustinus on the island of Capri in the last century BC. The first recorded life-sized reconstruction of a prehistoric animal based on a fossil tooth was made by Phlegon of Tralles some time around the end of the first century AD.

She relates that much Greek and Roman literature and art, which is linked to ancient myths, is based on fossils, showing that the ancient Greeks and Romans:

1. Recognized the organic nature of large fossil bones they found, and tried to visualize the origin, appearance, and behaviour of the creatures;
2. Realized that the bones were significantly older than the ones they usually found and therefore decided that they belonged to creatures that must have lived before humans were around to record their existence. This meant that such creatures no longer existed and they attributed their demise to local natural events.
3. Also attempted to recreate the animal's skeletons from the fossils they found.

Mayor reports that existing fragments of Anaximander of Mitetus writing from the 7th to 6th century BC indicate that he thought that the Sun's heat interacted with a primal ooze to generate sea creatures. He also thought that some of these developed via a "chrysalis" stage into the first humans that lived on the Earth's surface. Mayor interprets this as a recognition that he thought there was a progression of life, and that humans had adapted to changes in their environment in order to survive.

In the 4th century BC, Aristotle recognized that fossil shells found in sedimentary rocks were similar to existing seashells: from this he decided that the relative positions of the land and the sea must have varied in the past. Mayor points out that although Aristotle and other natural philosophers of the period managed to devise classifications for all living animals, they were tied to the idea of the "fixity of species", in other words they thought that species do not come, go, or change.

There is evidence that the ancient Egyptians used fossil logs for roadways in the deserts, but there is little evidence to suggest what they thought about their origins, whereas some Greeks appear to have appreciated the true nature of fossils. At around 600 BC Xenophanes, a Greek philosopher, is said to have observed the "impressions of small fishes" in rocks at Paros. He also found marine shells in the local mountains and found the impressions of fossil fish in deep mine shafts. Pythagoras and Xanthus of Sardis thought that shells found on the mountains were an indication that the mountains were at one time under the sea. Similarly, Herodotus, a 5th-century historian (Chapter 1), found marine shells inland in Egypt that he felt had been left by the sea.

According to Mayor, Lucretius, a Roman philosopher in the 1st century BC, "provided the clearest expression of extinction and 'survival of the fittest' in ancient literature", when he wrote that:

> Everything is transformed by Nature and forced into new paths. One thing dwindles ... Another waxes strong. In those days [the distant past], many species must have died out altogether and failed to multiply. Every species that you now see drawing the breath of life has been preserved from the beginning of the world by cunning, prowess, or speed. [Those] without natural assets fell prey to others, entangled in the fatal toils of their own being, until Nature brought their entire species to extinction.

Gaius Plinius Secundus – better known as Pliny the Elder (AD 23–79) – was an author, natural philosopher, and a naval and military commander who was killed trying to rescue several of his friends during the infamous eruption at Vesuvius. He agreed with the popular belief that fossils were supernatural. He referred to there being a "mineral ivory found in the ground" and that there were "bones growing within the earth". He mentioned various fossils, which he described as being "in the manner of a shell" or "like unto a sponge". He used the term "*Glossopetrae*" to describe fossils that "resembleth a man's tongue, and groweth not upon the ground, but in the eclipse of the moon falleth from heaven, and is thought by the magitians to be very necessary for pandors and those that court fair women". (The term *Glossopetrae* was used well into the 19th century, for what were eventually identified as fossil sharks' teeth.) Pliny also thought that jet was "not much different from the nature of wood", but was more interested in its uses than its origins, as the smoke from it "discovereth the falling sickness, and betrayeth whether a young damsel be a maid or no". Use of jet was particularly common on the Island of Malta. According to a Maltese tradition:

> These objects were miraculously created following the landing of Paul the Apostle after his ship was wrecked. Bitten by a Viper, Paul was angered but not poisoned. He placed a curse on all snakes thereabouts, whereupon their teeth turned to stone.

Although this may seem an extremely odd or even stupid idea now, very little changed from these views for nearly ten centuries. Throughout this period, fossils were used as evidence for the Genesis Flood. For example, Tertullian (155–222) – one of the early Christian church leaders – wrote that "to this day, sea conchs and tritons' shells are found as strangers on the mountains, desiring to prove to Plato that the heights have once flowed with water".

7.2.1 The Medieval view

Medieval thoughts about fossils centred on them being the product of "plastic flow" or "formative virtue" generated within the Earth. This idea was thought to be based on Aristotle's theory that living organisms were spontaneously generated by "vis plastica" and those fossils, being the shapes of animals without life, were the unsuccessful efforts of that force. Aristotle's ideas were developed by Avicenna (981–1037), an influential Islamic philosopher and scientist who wrote about medicine, astronomy, geometry, arithmetic, and music. He thought that the mountains might have been formed by two different forces; either by the effects of uplift caused by large earthquakes or by erosion caused by water cutting large valleys through high plateaus. He decided that the latter was most likely due to the presence of aquatic fossils.

Discussion point

Many Renaissance thinkers viewed fossils as the products of fantastic and mysterious forces operating within rocks, although there were some who considered them to be related to living organisms.

Albertus Magun (1193–1280), a Dominican scholar who was interested in alchemy, stated in his *Book of Minerals* that "certain stones having within and without the figures of animals", had been entirely changed "into stones, and especially into salty stones". He appears to have recognized the process of fossilization, whereby bones, shells, etc. are replaced by sediments, which maintain the form of the original but have been turned to rock.

Leonardo da Vinci (1452–1519), the natural philosopher, artist, and inventor, recognized the antiquity of shells and other fossils. He also recognized that *Glossopetrae* (see above), also known as "tongue-stones", were sharks' teeth. He also acknowledged the organic origins of fossils and discounted the Flood as the mechanism for their presence. He questioned why fossils were not found lying on the surface, rather than buried in rocks, if the widely held view that they had been deposited by a universal flood that covered all the mountains was true. Da Vinci also thought that, due to their weight, fossils would settle in the water rapidly during the deluge and could therefore not be deposited on the tops of the mountains after all.

Martin Luther (1483–1546) – the German monk and theologian – said of petrified wood, "I doubt that we have an indication of the Flood in the wood hardened absolutely into stone, which one often finds in places where there are mines".

Bernard Palissy (1510–1589) – a glass painter, surveyor, and potter – was famous for covering his pots in life-sized replicas of amphibians, reptiles, insects, and plants. He said that because there were:

> … rocks filled with shells, even on the summits of high mountains, you must not think that these shells were formed, as some say, because Nature amuses itself with making something new. When I closely examined the shape of the rocks, I found that none of them can take the shape of a shell or other animal if the animal itself has not built its shape.

In other words, he recognized the true origin of fossils.

In his *Britannia*, which was first published in 1586, William Camden (1551–1623) – an antiquarian and historian – referred to ammonites found in Gloucestershire as "little sporting miracles of Nature". He found round stones in Yorkshire, which "if you break them you find within stony serpents, wreathed up in cycles, but generally without heads (i.e. like the Whitby 'snake-stones')".

Discussion point

Palmer reports "that it was not until around 1557 that the first picture of a fossil or, at least what is generally claimed to be a fossil, was published", although Mayor proposes that many ancient paintings and carvings often include very good, relatively detailed representations of fossils.

7.2.2 *The 17th- and 18th- century view*

The 17th century saw an explosion in collecting natural objects, including minerals and "objects resembling living things", but it was still generally thought that fossils were of no particular historic importance.

Robert Plot (1640–1696), a naturalist and the first Professor of Chemistry at the University of Oxford, wrote in the *Natural History of Oxfordshire* that "stones in the form of shellfish" were "naturally produced by some extraordinary plastic virtue latent in the earth or quarries where they were found." He also thought that ammonites were formed by "two salts shooting different ways, which by thwarting one another make a helical figure".

A different theory was proposed by Elias Camerarius (1673–1734), a Professor of Medicine at the University of Tübingen, who wrote a history of epidemic fever. He said that the "seeds" of fossils were universally diffused throughout the ground and were capable of developing into their peculiar forms "by the regular increment of their particles".

Edward Lloyd (1660–1709), author and Keeper of the Ashmolean Museum in Oxford – who was also a botanist, geologist, and an antiquarian – wrote that fossils developed from "moist seed-bearing vapours" that rose from the sea and penetrated the Earth, possibly with the rain.

Others included Nicholas Steno (Chapters 1 and 2) who, in 1666, noted the similarities between fossil teeth *Glossopetrae* and present-day teeth from a white shark that he had dissected for the Grand Duke Ferdinand II. He published his findings, but due to its radical nature, these generated little interest until the 18th century, when they helped change the way in which fossils were viewed. It is interesting to note that, following the death of the Grand Duke and Steno's conversion from Protestantism to Catholicism, he gradually gave up science and became a priest. He eventually became a Bishop who covered northern and western Germany, Denmark, and Norway.

At about the same time that Steno was publishing his *De Solido*, an Italian named Agostino Scilla (1629–1700), a painter, palaeontologist, geologist, and a pioneer in the study of fossils, published *La Vana Speculazione Disingannata dal Senso* ("Vain Speculation Undeceived by Sense") in 1670. He pointed out that although he did not know how shells got onto the mountains, "what looked like seashells in the hills really were seashells".

Discussion point

Cutter points out in his book, *The Seashell on the Mountaintop: A Story of Science, Sainthood, and the Humble Genius who Discovered a New History of the Earth*, that by the time that Oldenburg's translation of *De solido* was published in 1671, the battle lines between the Church and science had already been drawn.

Martin Lister (1638–1712), a member of the Royal Society as well as a naturalist and physician, pointed out that fossil ammonites superficially resemble seashells, but bore no resemblance to any living molluscs. He was unconvinced that fossils were the remains of animals, and believed that they were imitations produced in the rocks by unknown forces.

Lister, who had met Steno, commented that although:

> … he did not doubt Steno that in some Mediterranean countries seashells might very well be found 'promiscuously included in Rocks or Earth', particularly along the coast … 'But for our English inland Quarries, which also abound with infinite number and great varieties of shells, I am apt to think, that there is no such matter as Petrifying of Shells in the business'.

He made two points that he considered proved that fossils were not real shells. Firstly, they were made of materials that were completely different to living shells, and that the fossil shells were made of the same materials as the rocks in which they were formed. Secondly, he pointed out that different quarries contain different types or species of shells, which are unlike those living on "land, salt, or fresh water".

Discussion point

Cutter makes an interesting observation with regard to the nature of fossils and their relationship to rock strata:

> So long as authorities such as Lister and Bounanni [A Jesuit priest in Rome] denied the biological reality of fossil seashells, and so long as the theories of plastic nature held sway in England and Rome, there was no reason for anyone to try. Steno's principle of superposition, sensible as it was, was irrelevant if strata weren't beds of sediment. And if the fossil seashells weren't seashells, there was no reason to believe that they were.

From the 1690s, naturalists began to struggle with the nature of fossils, asking questions such as:

Were they organic or inorganic, as they appeared to have different origins?

They knew that minerals "grew" with geometric forms and asked whether rocks and minerals were also able to grow in animal or vegetable forms?

They also noted that many fossils were unrecognizable as living animals. At the time, there was no known process by which the remains of animals or vegetation could be buried so deep in the ground, so high up mountains and hills, or so far away from the sea.

They also questioned why some animals could apparently be turned to fossils, whilst others could not?

Moreover, how could Nature have produced fragments of bodies rather than the whole thing?

As Porter notes, people were "moving away from Classical and Renaissance philosophies of natural history towards mechanical philosophy", which was then strengthened by Newtonianism.

Robert Hooke (Chapter 1), the son of a clergyman, was born on the Isle of Wight and became one of the most important scientists in the country. He was also largely responsible for the layout of the City of London after the Great Fire of 1666, when he was appointed City Surveyor under Sir Christopher Wren – a job he did whilst he was still the Curator of Experiments for the Royal Society.

In the latter role, he was responsible for all of the experiments conducted at the Royal Society, during which he discovered a number of the fundamental properties of earthquakes. He also invented a compound microscope, which he used to compare the structure of wood with examples of fossilized wood; from this, he decided that fossil wood was the remnants of living trees. He recognized two different kinds of fossils: one included wood and bones, and the other type were replacements of the original material. He also rejected the idea that the sediments they were found in originated from the Flood. He identified Earth movements as causing the elevation of rocks – "many parts which have been sea are now land" – and the existence of extinctions due to catastrophes, but he could not envisage a long geological time scale. He also suggested using fossils as a chronological index; the extinction of species; variation and progression due to changed conditions; climate changes inferred by fossils; and that modern forms of organisms might have evolved from those found as fossils.

John Woodward (1665 1728) – a member of the Royal Society, sometimes known as the "Grand Protector of the Universal Deluge" – extended Steno's ideas and linked them to the Flood. Woodward found his first fossil when he was 25 years old and went on to collect and own one of the largest fossil collections in London. In fact, he boasted that after only four years of collecting, he had covered almost the entire country. In a book entitled, *An Essay Towards a Natural History of the Earth*, published in 1695, he noted that particular fossils were found in particular rocks and he attributed this to the effects of the Flood – a view that is frequently still used by modern-day creationists.

An error was pointed out at the time by John Ray (1627–1705), the son of a blacksmith who went to the University of Cambridge, later becoming an expert in languages, mathematics, and natural science and later still becoming an Anglican Priest and an eminent naturalist. He believed in creation but recognized that if ammonites were extinct, it meant that they had left a gap in the plan of creation. He also doubted that the Flood could account for their distribution. Ray was concerned that fossils were being recovered from all over the country – including those of elephants in Oxfordshire, hippopotamuses in London, and corals from a number of different sites spread across the land.

Others, such as the French author François-Marie Arouet (better known as Voltaire (1694–1778)), thought that the fossils found in the Alps were "simply shells left by pilgrims on their way to Saint James of Compostela in Spain".

Discussion point

By the end of the 17th century, the Flood was no longer considered the principal mechanism for the change and distribution of fossils. However, the amount of information available and the level of understanding about what fossils were and what they represented, was at the time insufficient to settle the argument.

In the 18th century, people were becoming more willing to challenge the authority of the Church, but no one would go along with Hooke's *Unstable Earth*. It was during this period that universities and museums began to act as centres of important scientific collections. The number of scientists and others studying fossils continued to increase and, as a consequence of their studies, many scientists began to change their minds with regard to the origins of fossils. Many felt that they had to be the remains of living creatures and that the Flood could therefore not account for their complex distribution.

An interesting situation arose in Germany, involving Professor Johann Bartholomew Adam Beringer (1667–1740), a Professor of Medicine at the University of Wurzburg. Several of his colleagues carved blocks of limestone into the shapes of lizards, frogs, and spiders, hiding them in the locations on Mount Eibelstadt where Beringer usually went to find fossils. Beringer published descriptions of these "fossils" in a book titled

Lithographiae Wirceburgensis in 1726. Although these "fossils" were a hoax, Beringer was very careful in the way in which he studied them. His book included not only his own interpretations, but he also included other people's explanations. He thought that some might be the fossils of dead animals, but others were the "capricious fabrications of God" that had been made to test humans, as some included the name of God written in different languages. Even though some clearly showed evidence of being chiselled (it is estimated that it took up to six hours to produce some of them), he went ahead and published his descriptions. It is said that when he discovered that the "fossils" were fakes, he tried to buy up and destroy every copy of the book and almost made himself bankrupt in the process. The episode is usually known under the title Beringer's "lying stones".

Buffon (Chapters 1 and 5), in his *Histoire Naturelle*, wanted to create a comprehensive account of the Earth and its inhabitants. He thought that fossils represented extinct early life forms that had died when the warm oceans in which they lived cooled down. He also used the fossils of woolly mammoths found in Siberia as evidence of this, as the Earth's surface also cooled down.

7.2.3 The 19th-century view

As we have seen in Chapters 1, 5 and 6, Georges Cuvier worked on marine invertebrates of the Tertiary Period in the Paris Basin, and is recognized as one of the people who established biostratigraphy (Chapter 2) as a method of dating and ordering rocks. In 1830, Cuvier and Geoffroy St Hilaire (1772–1844) – a French naturalist and friend of Lamarck's – took part in a famous debate at the Royal Academy of Science in Paris. St Hilaire thought that all organisms had developed with differences in the same basic form, resulting in his idea of "unity of plan", whereas Cuvier thought that differences in form were based on the separate body forms, as outlined in Chapter 6.

Discussion point

This discussion has often been portrayed as an opening round in the debates on evolution, but it was really a discussion about the number of different basic body forms with which life had developed. Remember, at the time many naturalists felt that fossils were still evidence of the Flood.

Agassiz (Chapter 5), who became Professor of Natural History at the University of Harvard and founded the Museum of Comparative Zoology, believed that organisms became more complex and better suited to their environments over time, through a series of independent acts of creation by a "Supreme Being".

Discussion point

Fossils appeared to show gradual changes and therefore became one of the major proofs of evolution. Despite the apparent randomness of the preservation and distribution, there were usually enough fossils to provide a relatively coherent story.

A fossil can be compared to other fossils and/or living animals, thus providing us with a basis for interpretation. It is the way in which they are interpreted that varies and can be called into dispute. We see the products of evolution, not the links between them.

Generally, the creationist view of fossils is that although some rocks are rich in fossils, it is hard to reconcile this with the lack of fossils being formed today: in other words, you do not find fossils forming today that are clearly made of rock.

The 19th century signified a period of change with regard to the study of geology. In her book *The Dinosaur Hunters: A True Story of Scientific Rivalry and the Discovery of the Prehistoric World*, Deborah Cadbury says that in a period of perhaps 12 years during the early 1800s, Buckland's "undergroundology" had blossomed into the "Queen of Science"; British geologists had effectively mapped and ordered the succession of rocks previously considered to be Secondary and Tertiary in age and had begun to think of them as representing a record of significant changes in the Earth's history. This was accompanied by a series of discoveries of incredible marine fossils, such as the Ichthyosaurs and Plesiosaurs found by Mary Anning on the Dorset coast, and the gigantic terrestrial reptiles found by Buckland and Mantell. These helped lead to a change in emphasis from concentrating on detailed studies of rock units to attempting to understand fossils and the worlds in which they lived (Fig. 7.2).

7.2.4 Mantell versus Owen

Many characters were involved in the discovery, identification, and classification of the dinosaurs but the battle between two in particular, Gideon

Fig. 7.2 **Two views of the Jurassic Lias rocks exposed at the foot of Black Ven, Dorset**

Mantell and Richard Owen, shows the best and worst sides of scientific endeavour. Cadbury's book provides a very interesting and easy-to-read account of the twists and turns, the winners and losers, the good guys and the bad guys involved in the early discovery of what are arguably the most iconic prehistoric animals that ever lived on Earth.

Cadbury begins her story with the rivalry between Owen and Mantell, by putting it into context within a wider British and European setting. She notes that whilst the French had Cuvier and the Swiss had Agassiz, the British considered themselves at a disadvantage, particularly as they were discovering such remarkable fossils. The power brokers in the British Association for the Advancement of Science decided that they needed someone to interpret and classify these fossils "before they fell into the hands of foreigners". They realized that such a person would have to have the same scientific standing of Cuvier, and they chose Owen rather Mantell to fill that role.

But why did they make that decision? Who were these two characters, and what were their merits for such a worthy position?

Gideon Algernon Mantell (1790–1852) (Fig. 7.3), born in Lewes in Sussex; the son of a shoemaker, became a doctor, but his passion from childhood was geology. He was particularly interested in the "Ichthyosaurus" fossils that Mary Anning had discovered in the rocks at Black Ven on the Dorset coast (see below). Mantell wanted to know how they compared to the bones and teeth that he had found at Cuckfield, Sussex in 1820. The fossils at Cuckfield were the first terrestrial-living ones found from the Cretaceous in England, whereas Anning's Dorset samples were of marine origin from the Lower Jurassic. He sent some of the Cuckfield fossils to George Greenough to identify. Greenough compared Mantell's fossil bones and teeth with the drawings produced by William Conybeare, who had undertaken a detailed study of Mary Anning's Ichthyosaurus for the Geological Society. It should be noted that Anning also found the first

Fig. 7.3 **Gideon Algernon Mantell**

plesiosaur and the first British pterosaur; in fact, it was Conybeare who found a skull that he thought belonged to a plesiosaur, which enabled Anning to complete her specimen. At the time, Cuvier thought that it was a hoax but later, more complete finds proved that Anning and Conybeare were right.

Whilst Mantell was struggling to understand what he had found, geological folklore has it that his wife Mary found a giant tooth by the side of a road whilst waiting for her husband. This did not fit with one of the original interpretations that it belonged to a crocodile. It is interesting to note that Mantell was self-taught, and did not have the backing of a university or prestigious society and was deemed unqualified and in no position to interpret or classify such a remarkable beast. After all, he had proposed that the bones indicated that the creature was between 30 and 40 feet (9–12 m) in length, and no animal that size had ever been found.

In 1818, Cuvier visited the University of Oxford to look at the giant bones that had been found in the Jurassic rocks in the area at Stonesfield. Cuvier identified these fossils, which were in the care of William Buckland, as belonging to a giant reptile around 40 feet (9 m) long. In 1821, Charles Lyell visited Mantell in Sussex and told him about Buckland's fossils, some of which he subsequently had sent to Mantell for inspection. At about the same time, it appears that Mantell heard from Lyell that Buckland was going to publish a paper on the Stonesfield reptile fossils. This left Mantell in a difficult and vulnerable position. The teeth he had found indicated that his was also a large carnivorous reptile similar to the ones Buckland was about to publish. This would mean that Buckland's interpretation would be adopted. Buckland, being a highly regarded academic, was also in a better position to describe and publish details of his reptile, a situation that would result in Mantell losing the recognition he felt he deserved. He therefore decided that he would have to get his finding published first.

As Christopher McGowan puts it in his book, *The Dragon Seekers: The Discovery of Dinosaurs during the Prelude to Darwin*, "even in our modern world of wonders, we still marvel at the latest dinosaur discoveries, so imagine what it must have been like back in Buckland's time". Michael Freeman (in *Victorians and the Prehistoric: Tracks to a Lost World*) helps put these discoveries into context within the revolutionary changes taking place prior to and during the early decades of the Victorian era. He writes:

> There are senses in which the dinosaur stood proxy for a whole range of currents within early and mid-Victorian thought. Just as the steam railroad, for all its terrible monstrosity and apocalyptic imagery, was a central emblem of the march of engineering sciences, so the unfolding underground worlds of extinct reptiles was central to the march of Earth science.

When Mantell took his findings to the Geological Society in London, his ideas were dismissed, not because of his description, but because the members of the society thought that he had placed the geological sequence in the wrong part of the geological time scale. They thought that the sequence from which he had found his fossils belonged to the Tertiary rather than the Cretaceous Period. As Cadbury observes:

> Mantell's uphill struggle to get his ideas accepted by the experts was not unique. One amateur geologist, Robert Bakewell, was not allowed to join the Geological Society although he wrote a popular book, *Introduction to Geology*, wrote frankly about the difficulties.

He realized that the class system that operated within the established scientific societies meant that anyone who lived outside the major cities was incapable of achieving anything "important for Science". William Smith held similar views, observing that "the theory of geology was in possession of one class of men and the practice in another". Eventually, in 1824, Cuvier confirmed that Mantell's fossils belonged to an unknown herbivorous reptile, whereas Buckland's Oxford fossils belonged to that of a carnivorous reptile. On advice from William Conybeare, Mantell named his fossil *Iguanodon* and finally gained the fame he deserved. He eventually became a member of the Royal Society in 1825.

Discussion point

McGowan says of the Geological Society of London:

> This learned society was founded in 1807 to acquaint geologists with each other for "stimulating their zeal", and to communicate new facts. Buckland and Conybeare both joined in 1811; Lyell joined in 1819, Mantell in 1820, and Hawkins in 1832. Darwin became a member in 1836 … Owen joined a year later. Mary Anning, being a woman, would not have been allowed to join even if she had wanted to, but she was made an honorary member after her death.

A number of people were involved in collecting these fascinating fossils, including Thomas Hawkins (1810–1889) – a private collector who lived in Somerset – who spent his inheritance trying to obtain the best specimens from Lyme Regis. It is reported that on one occasion he even paid for someone to blow up a section of the cliff in the hope of exposing an ichthyosaur. He put together a huge collection of fossils, including some of the largest and best examples of ichthyosaurs, which he eventually sold to the British Museum.

Richard Owen (1804–1892) (Fig. 7.4), who was born in Lancaster, studied medicine at the University of Edinburgh and Bart's Hospital in London, and then specialized in anatomical research at the Royal College

Fig. 7.4 **Richard Owen**

of Surgeons, where he became the Hunterian Professor. His work is said to have covered every type of animal, including extinct reptiles. As we saw in Chapter 6, Owen was asked by Henslow and Darwin to work on some of the fossils that Darwin had brought back from his voyage on the *Beagle*.

Owen won the backing of the three-man British Association for the Advancement of Science committee to write a *Report on the Present State of Knowledge of the Fossil Reptiles of Great Britain*. Interestingly, the three committee members were Greenough, Lyell, and William Cliff, Owen's father-in-law. The first part of the report looked at Mary Anning's ichthyosaurs and plesiosaurs, many of which had been covered by Conybeare. The second part covered the saurians, which included the fossils found by Mantell.

During his travels, Owen met William Saull, who had been collecting fossils from Tilgate Forest, the site where Mantell had recovered most of his fossils. Saull let Owen have complete access to his own extensive collection on the promise that he would have credit for his finds. Owen also had access to Mantell's collection, which was then held in London. Unfortunately, this meant that all of Mantell's hard work had fallen into the hands of the person most determined to see his demise. At last "Owen began to 'reap the rich harvest' of which the leaders of science such as Sir Philip Egerton had sown the seeds". Mantell's downfall at the hands of Owen affected not only his life's work but also his marriage and his finances, a situation that led to him having to sell his fossil collection.

Owen's report included dividing saurians into four separate groups:

1. *The Enaliosauria* – a name proposed by Conybeare – which included ichthyosaurs and plesiosaurs that had lizard-like characteristics;
2. *The Crocodilian Sauria*, which had been identified by Cuvier;
3. *The Pterodactyls*, or flying lizards;
4. *The Lacertians*, which included Iguanodons, Hylaeosaurus, and Meglaosaurus.

Owen thought that the Lacertians contained "Reptilian type of structures", that "made the nearest approach to mammals", and came up with the idea of combining the terms "*deinos*" – meaning terrible or fearfully great – and "*sauros*" – meaning lizard, into the name *Dinosauria*.

He wrote:

> The combination of such characters, some, such as the sacral ones, altogether peculiar among Reptiles, others borrowed, as it were, from groups now distinct from each other, and all manifested by creatures far surpassing in size the largest of existing reptiles, will, it is presumed, be deemed sufficient ground for establishing a distinct tribe or suborder of Saurian Reptiles for which I would propose the name of Dinosauria.

As Cadbury puts it:

> In these few words, as he quietly redrafted his paper on that fateful afternoon, Richard Owen sealed the fate of Gideon Mantell. In this giant conceptual leap, as he defined the characteristics of his Dinosauria, he cast a spotlight on his brilliance at interpreting the fossil record.

It was clear that, although Mantell knew of the existence of fossil reptiles for a long time, Owen's invention of the term "dinosaur" meant that it would be he, and not Mantell, who would be credited with their discovery. Owen also appears to have backdated some of his ideas to make it look at though he had them before some of his rivals.

Owen attacked the early evolutionists as he thought the fossils showed that "there was no graduation of one form into another". He also took a swipe at Mantell, arguing that there was no similarity between fossil and modern iguanas, and even going as far as to say that the name *Iguanodon* was therefore "inappropriate". Mantell thought that Owen had used his descriptions of the fossils features as though Owen himself had described them – in other words, he had plagiarized Mantell's work. Cadbury records that Mantell thought his behaviour showed "unworthy piracy and ingratitude". In 1846, Owen published a paper on belemnites in which he completely ignored the work of Chaning Peace, which had been presented to the Geological Society four years earlier. This omission was described as "an evil of no slight magnitude in the progress of scientific research" by Edward Charlesworth, editor of the *London Geological Journal*. Cadbury also points out that Owen introduced the term "dinosaur" into his report **after** he had presented his findings to the BAAS in August 1841.

In 1848, Mantell was sent part of a jawbone from the Tilgate Forest quarry, which provided the evidence he had been looking for. He was invited to present his findings to the Royal Society but, after he had finished speaking, Owen announced that a similar, but smaller jawbone had already been found. Mantell had also been looking at backbone vertebrae with which he hoped to be able to reconstruct an Iguanodon spine, in order to be able to determine how long the animal had been when it was alive. He thought that the vertebrae that Owen had identified as belonging to several different reptiles were actually all from the same one, and he presented his findings at the Royal Society in 1849, even though Owen had tried to dissuade him from doing so. Mantell was the first person to recognize that the Iguanodon's front legs were significantly smaller than its back legs.

This, together with other finds, meant that Mantell was "challenging Owen's supremacy in the field of dinosaurs". In 1849, Mantell presented another paper to the Royal Society, in which he described a new dinosaur, the Pelorsaurus, the first of the sauropod (or "lizard foot") fossils to be found. A short while later he also identified another sauropod that Owen had incorrectly thought was related to crocodiles. Cadbury tells us that:

> With this success, towards the end of 1849 Gideon Mantell's name was proposed once more for the prestigious Royal Medal of the Royal Society. But he learned that the committee passed over his paper on Iguanodon because of Owen's disparaging remarks.

In fact, the committee and Council of the Royal Society had met on three occasions to decide the matter. Each time, Richard Owen did everything that he could "to prevent the award being made to Mantell". At a fourth meeting, Lyell and Buckland spoke up for Mantell, and he was at last awarded the medal.

The battle between the two continued, and when Owen wanted to publish a paper on British reptiles, he implied that some of Mantell's work was his own. Mantell managed to expose this falsity at a council meeting of the Royal Society, and Owen was forced to apologize. By this time, it appears that Owen was arguing with almost everyone. He continued to claim other people's work as his own and eventually even some of his most loyal supporters began to turn against him.

When the Great Exhibition was moved from its original location in Hyde Park to Crystal Palace, Mantell was asked to oversee the production of life-sized replicas of the dinosaurs. Unfortunately, he felt that due to his poor health he had to decline the work. Shortly afterwards he fell over and, after taking opiates to relieve the pain, he overdosed and died. An anonymous obituary appeared in the *Literary Gazette*, in which he was described as an inadequate scientist. Cadbury records that "even the discovery of Iguanodon was taken from him". According to the obituary:

> To Cuvier we owe the first recognition of its reptile character, to Clift the first perception of the resemblance of its teeth to those of the Iguana, to Conybeare its name, and to Owen its true affinities among reptiles, and the correction of errors respecting its bulk and alleged horn.

Cadbury and other authors add that all the achievements and international acclaim appear to have fed Owen's ego, leading him to be even more vicious to anyone he saw as a rival, particularly Mantell, whose reputation he wanted to destroy completely. Most people thought that Owen had written the obituary and were shocked by this final attack. As Palmer writes in his excellent book *Earth Time: Exploring the Deep Past from Victorian England to the Grand Canyon*, "Owen seems to have been a thoroughly unpleasant man and few of his contemporaries had a good word for him, but nevertheless his anatomical talent does have to be acknowledged".

When Mantell had to decline the offer to oversee the design and construction of the dinosaurs for the Great Exhibition, Owen was put in charge, and it is his dinosaurs that still exist in the park at Crystal Palace (southeast London) today (Fig. 7.5). It is interesting to note that although the dinosaurs took centre-stage, the area of the park in which they were sited, according to Freeman, was designed as an "open-air museum". This included an artificial lake with a five-foot tidal range, a limestone waterfall (Fig. 7.5(e)), gently folded limestones complete with a small cave (Fig. 7.5(f)), backed by a wall complete with a "mock geological strata" (Fig. 7.5(g)). Owen even wrote a guidebook that helped turn the area into "a great lecture room on geology".

Even though the Biblical Flood had been largely discounted by then, Owen designed these dinosaurs (Figs 7.5(a–c)), based on the limited interpretations of the time, to be representatives of the primitive creatures destroyed by the Flood, an act that would also show that evolution did not exist. Remember that although he was asked to work on some of the fossils brought back by Darwin, he rejected Darwin's ideas of evolution.

In 1856, Owen was to be offered the new post of Superintendent of the Natural History Department of the British Museum in London. He proposed constructing a new museum building that to him would be a monument to God. One of the people to oppose his design was Thomas Henry Huxley, who we shall meet in the following sections.

By the time that the Natural History Museum in South Kensington was finished (at one time the Thames Embankment had been the preferred site), Owen was old and frail. Even so, it appears that he still managed to claim one more victory over Mantell. He dispersed some of Mantell's collection to a number of different museums around the country, so that it could never be viewed together in one place.

His battles with Darwin's supporters also took their toll on Owen's reputation, and after he died in 1892, Cadbury records that he was systematically written out of history by the Darwinians because of his opposition to Darwin's theory. As she puts it, "his personality was blackened, his treatment of rivals condemned" and "so complete was the assassination of his reputation that, within a few years, one Oxford professor dismissed him merely as a damned liar". But, as McGowan says, "while Owen may not have been the nicest of people, Mantell was not the easiest of men. His [Mantell's] disputes with others were not uncommon – from quarrymen and collectors, to employees and family members".

Discussion point

Do you think that Owen's actions reflect examples of the victor taking the spoils of war?

Fig. 7.5 **Dinosaurs at Crystal Palace and the Natural History Museum, London. Three views (a–c) of the Owen's dinosaurs preserved at Crystal Palace, south London, which was originally built for the Great Exhibition. These include a Plesiosaurus (lower centre model in photograph (a)) and two Teleosaurus. Photograph (b) shows two Iguanodons. Photograph (c) provides a general view across the "Primary and Secondary Island", with several examples of Labyrinthodon in the foreground. Photograph (d) shows the famous Diplodocus skeleton that greets visitors to the Natural History Museum in London. Photographs e–g present the other geological phenomena, including a waterfall (e), a folded limestone sequence complete with a cave (f), and the "mock geological strata" (g)**

Fig. 7.6 **Mary Anning**

This was not the only instance of dishonesty and rivalry at the time. We have already heard of Mary Anning (Fig. 7.6), who became world famous for collecting fossils from the beaches and cliffs around Lyme Regis in Dorset. She and her mother owned a small shop close to the seafront in which they sold fossils. Many people bought their samples including, as Palmer reports, the Philpot sisters, who eventually donated their collection to the University of Oxford Natural History Collection. Indeed, as Patricia Piece reveals in the book *Jurassic Mary: Mary Anning and the Primeval Monsters*, many of the most famous geologists of the time visited and brought fossils from the Annings. According to Cadbury, Mary wrote to a friend of hers saying, "these men of learning have sucked her brains, and made a great deal by publishing works, of which she furnished the contents, while she derived none of the advantages". As we saw earlier, her work was not really acknowledged until after her death. Her relationships with the "men of science" was not all one-sided, however, as William Buckland in particular went to great efforts to try to raise funds for the Annings when they fell into financial trouble, and Adam and Charlotte Sedgwick were both very good friends.

Discussion point

It is interesting to note that 50 years after Mary Anning's death, Terry Sullivan wrote the famous tongue twister, "she sells seashells, by the sea-shore", based on her life.

Another example of dishonesty at the time is Sir Everard Home (1756–1832), an assistant to his famous brother-in-law, John Hunter, who was considered to be the "father of modern surgery". It is reported that after Hunter died of a heart attack, Home published his brother-in-law's work under his own name. When this was discovered, he was described as "not only incompetent, but also a fraud". Whilst Buckland was working with Cuvier on Mary Anning's discoveries, Home rushed his own interpretation into print: in this, he concluded that the fossil was a crocodile but its teeth indicated that it could not be a reptile. Later he thought that it could be an "aquatic bird" or even a fish, and decided to call it a *Proteosaurus* ("proteus-lizard"). However, it had already been named *Ichthyosuarus* by Charles Konig at the British Museum.

Discussion point

What do you think drives scientists to be dishonest, and could the same thing could happen today?

Philip Manning's book, *Grave Secrets of Dinosaurs*, presents a very good, concise account that shows how "over enthusiasm" can still affect scientific judgement nowadays, particularly with regard to dinosaurs. He explains that:

> Dinosaur soft-tissue fossils, as a class, have become a kind of Holy Grail to many fossil hunters and researchers – a goal so alluring that the search has caused errors in judgement even among high-ranking, well-regarded professionals. When people want something badly enough, they often develop the ability to see what they want to see, regardless of what is actually there. Thus, along with legitimate finds of great value and scientific interest, the history of dinosaur soft-tissue fossils is also spotted with eagerly promoted miracle finds that, on closer inspection, turn out to be mirages.

> In most cases, these are not hoaxes or deliberate misrepresentations. Rather, emotional excitement at the spectacular possibilities, or a thirst for fame and recognition, has overcome cool-headed and clear-sighted analysis. This background of occasional, but prominent, mirages in dinosaur soft-tissue fossils serves as a set of cautionary tales for anyone encountering what appears to be a "find of the century".

His book contains a very good account of the over-enthusiasm induced by the remarkable fossils coming out of the Liaoning shale beds in China. He relates the story of Stephen and Sylvia Czerkas and their purchase of an apparent dromaeosaurid dinosaur for their museum in Utah, and the subsequent publicity and eventual discovery that the fossil was a compilation of at least two separate animals.

Discussion point

For scientists, this clearly shows the importance of careful, peer-reviewed studies, in which independent professionals scrutinize other people's findings. It shows that when scientists make claims, the "system" of peer review is self-regulating.

Going back to the time when Buckland, Anning, and Cuvier's discoveries and controversies were being made in the UK, dinosaur hunting had begun in earnest in North America. It is sometimes easy to forget that whilst we in the UK and much of Europe were experiencing rather gentle lifestyles, much of America was still being discovered. Manning brings home this point when he relates that, following Britain's so-called "heroic age of geology" in the first half of the 19th century, America was involved in a fossil "bone rush". During the cowboy era of the "Wild West", people such as Edward Drinker Cope (Fig. 7.7) and Othniel Charles Marsh (Fig. 7.8), were fighting their own feud as they both tried to outdo each other with their finds. The feud became so intense that "their men in the field spied upon one another, and rival teams reportedly came to blows over fossils at times". This did, however, provide unexpected benefits, for "with extensive financial backing available to both men, they drove scouts and prospectors and excavators who returned tremendous benefits to the science of paleontology and the study of dinosaurs in particular".

7.3 Famous Evolution versus Creation debates

Before we look at the historic and ongoing "battle" between evolution and creation, it is necessary to include the following diagrams which, in a simplistic way, provide in visual form, the basis of each view point.

Fig. 7.7 **Edward Drinker Cope**

Fig. 7.8 **Othniel Charles Marsh**

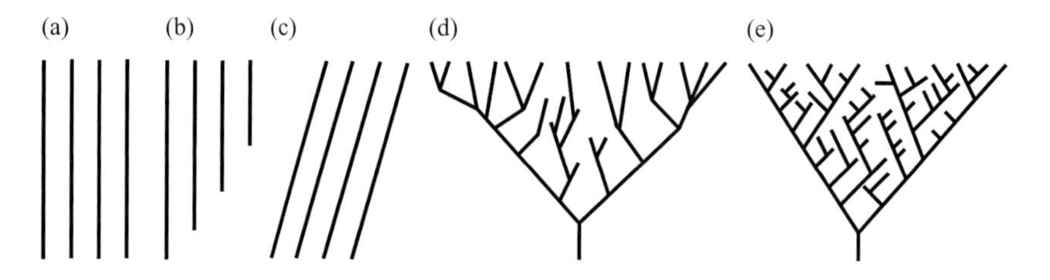

Fig. 7.9 **Visual representations of the different ideas of development of life**

Figure 7.9(a) represents the creation of all life forms, over a short period each in their own kind. Following their creation, they remain the same. Figure 7.9(b) indicates a similar "process" of creation, but with the introduction of different "kinds" at different times in Earth history. Figure 7.9(c) represents Lamarck's transformism, whereby animals adapt, over time, to changing environmental conditions, but remain the same species. Two different views of evolution have been included, the first (Fig. 7.9(d)) represents an evolutionary tree, and the second (Fig. 7.9(e)), is an evolutionary bush. The basic difference between these two models is that with the latter, numerous new species evolve but only exist for a comparatively short period.

As we have seen, the publication of Darwin's theory of evolution in *On The Origin of Species* generated widespread and often heated arguments between those who supported the theory of evolution and those who maintained that life developed purely as an act of God – many of these discussions still rage today.

Among these are three of the most famous debates about the difference between evolution and creation. They can be found in a number of different books, including Gould's *Bully for Brontosaurus*. The following are a brief overview of the arguments.

7.3.1 Huxley versus Wilberforce

Following the publication of Darwin's *On The Origin of Species* in November 1859, one of the most famous debates linked to the book occurred about six months later on Saturday, 30 June 1860. This was a debate between Thomas Henry Huxley (1825–1895) (Fig. 7.10) – a biologist and a friend of Darwin known as "Darwin's Bulldog", and Samuel Wilberforce (1805–1873) – the Lord Bishop of Oxford and the third son of William Wilberforce (the man who led the fight to abolish slavery). It took place at a meeting of the British Association for the Advancement of Science in the Zoological

Fig. 7.10 **Thomas Henry Huxley**

Museum at the University of Oxford, in front of approximately 700 people. In fact, so many people had turned up for the meeting that it had to be moved from its original location to the great library.

The debate was based around a presentation of "the intellectual development of Europe considered with reference to the views of Mr Darwin". It was a debate that everyone had specifically come to hear, and a debate that has been described as one of the "six legends of science", although it subsequently appears to be more of a myth than a legend. It has also been touted as the point at which science became autonomous from the Church. In the official version, it is said that Wilberforce spoke for half an hour, during which he savagely ridiculed Darwin and Huxley. He then turned to Huxley and asked him, "if it was through his Grandfather or his Grandmother that he claimed to be descended from an ape".

Huxley is then said to have replied that he would "feel no shame in having an ape as an ancestor, but that he would be ashamed of a brilliant man who plunged into scientific questions of which he knew nothing". The room is then said to have dissolved into uproar and the debate finished.

Why is this debate so famous and so important? It is portrayed as the first time that the world's attention was focused on one of the real issues of 19th century – the question of science versus religion. It is widely thought that Huxley "won the debate" and therefore science triumphed over religion.

Although this is the official version, which presents it as a clash between science and religion, it appears that the truth of the situation may be somewhat different. Even though this was an official meeting of the BAAS, and everyone who attended did so with expectations of an important debate, no records were taken of the meeting and so there is no official transcript of what was actually said. The official version was written by Darwin's son 25 years after the event. Darwin did not attend the meeting because he was ill, although it has been said that he may simply not have wanted to present the argument himself. One individual who was there later wrote that he and most other people thought that Wilberforce rather than Huxley had "won" the debate. A number of famous scientists were there, including Richard Owen – who is said to have "coached" Wilberforce so that he could take part in the debate. In fact, it appears that Wilberforce had hoped that someone else might oppose Darwin's views in the debate.

Wilberforce had already written a review of *On The Origin of Species*, in which he took a balanced approach to what it said. He made his arguments on scientific grounds and did not do so simply to uphold the Church's view. He did not argue that the theory was wrong, because of its implications for humans but because there was a lack of proof.

Wilberforce pointed out the following weaknesses in the theory:

1. That over the entire period of human history there had never been any evidence of the development of a new species. (We now know that has actually happened.)
2. That selective breeding produces changes, but when selectivity ends species revert back to their original type.
3. All hybrids result is sterility.
4. Because of a lack of evidence, he argued that Darwin's ideas should be regarded as a conjectural hypothesis rather than a well-established theory.

It appears that Darwin thought that Wilberforce's criticisms were fair and that he himself had highlighted some of the difficulties with his theory. During the debate, Wilberforce based his argument on his review, which he later published. It is quite probable that, as Wilberforce's knowledge and understanding of science was not particularly strong, Owen wrote much of the review – especially as we have seen that he disagreed with Darwin's views.

It is said that Huxley spoke only briefly and presented no details against Wilberforce's views. He too pointed out the lack of evidence in the geological record for changes in species and instead focused his remarks on the logic of Darwin's arguments.

The debate was followed by a summary by Joseph Dalton Hooker (1817–1911) – one of the most famous botanists of the 19th century and a member of Darwin's inner circle – in which he stated the botanical aspects of the theory. In fact, it was Hooker who said that Wilberforce had distorted and misunderstood Darwin's theory, i.e. that one species changed to another through transmutation rather than through the successive development of species by variation and natural selection. Hooker also said that he himself had opposed the ideas of evolution, but had changed his mind following his long study of the form and distribution of plants. It is said that Wilberforce did not respond to Hookers summary and the meeting closed.

The key point of the debate was that it led to a continued discussion about how science and religion fit together under the banner, "enlightenment versus reaction, truth versus dogma, and light versus darkness". One of the most important consequences of Darwin's book was that it affected the whole of biological thinking from that time onwards.

Christine Garwood, in her book *Flat Earth: The History of an Infamous Idea*, reports that Huxley was one of a group of young scientists who were determined to "wrest influence from the established intellectual elite. Determined to release science from its links with aristocratic interests and the Anglican Church, and win social and intellectual authority for themselves and their peers".

Discussion point

It has often been suggested that Wilberforce's comments were based around a single sentence, which effectively said that Darwin's theory might "throw light on the origin of man and his history" – but there is no other discussion in *On The Origin of Species* on the subject of humans.

7.3.2 Huxley versus Gladstone

The debate continued some years later between Huxley and William Ewart Gladstone, the famous Prime Minister. In 1885, Gladstone published an article titled the *Dawn of Creation and of Worship*, in which he tried to reconcile the Genesis story with geology to establish its scientific truth. Part of his argument was that at the time Genesis was written, the authors would not have had the knowledge or evidence to identify the order of events, to be able to compile their account. He therefore argued that they must have had the information revealed to them by God.

His main points were that the six days of creation were not six literal days, but six steps. He pointed out that the general order of the Genesis account broadly fitted with what was known of the early history of the Earth until the introduction of Darwinism.

He argued that the first four days were the cosmological events, and the biological events took place on days five and six. In the latter, he placed special emphasis on a four-fold sequence in the appearance of animals, i.e. water populations followed by the air populations on day five, and land populations followed by humans on day six.

In fact, if you look at Genesis, chapter 1, this is not strictly true, as it says in verse 11 that the Earth produced plants – i.e. a biological event – during day three.

The following month, Huxley countered Gladstone on four points:

1. That land animals appeared before flying creatures, even if you discount amphibians and reptiles as land creatures;
2. That on anatomical grounds, flying creatures had to have evolved from land creatures, because of the structures they required for flight;
3. That the geological record indicates that new species have appeared continuously throughout the geological record;
4. That plants appeared on the third day before animals, but in the geological record, animals appear first.

Gladstone only answered the third point, arguing that Genesis only refers to the sequence of their first appearances and not to the subsequent generation of new species, quoting the stone, bronze, and iron ages as examples. He argued that this division does not imply that stone tools, for example, were not made in the following periods.

Discussion point

The views of Stephen Jay Gould, an evolutionary biologist, are interesting.

Gould said he could not view the Genesis account as a story of linear addition but as differentiation: God created a chaotic formless total and then made divisions within it. He points out that this follows the rules of DNA coding as, for instance, the range of structures needed for flight are fairly few and there are only a few pathways in which something can evolve from a simple creature to a more complex one.

An interesting review of the Genesis version of creation appears in *Genesis Today: Genesis and the Questions of Science* by Ernest Lucas, a scientist and church minister. He points out that the first three days are concerned with shape and subdivision of the Earth from its "shapeless and empty" original form. The second three days are concerned with filling the Earth with creatures that are suitable to live in its different parts. He also outlines a number of different ways that the Genesis account has been viewed within the Church.

It is clear that in Genesis the Earth was formed, as Gould says, by continuous subdivision from an original in which nothing else was added; effectively, the Earth was a closed system (although we now consider it to be an open system).

Having read and heard many interpretation of Genesis, the Bible actually says nothing about the process of creation. This means that views on the way in which creation occurred are interpretation, and it is important to separate fact and evidence from interpretation.

There is also a very good discussion about different interpretations of the creations story in David Snoke's book, *A Biblical Case for an Old Earth.*

7.3.3 The abolition of the equal time laws in America

In 1925, a famous trial took place in Tennessee that has often been called the "Scopes Monkey Trial". This trial was effectively instigated to test a law passed in 1925, which forbade the teaching of any theory that denied divine creation in any state-funded education. There is also some evidence that the trial was staged to generate publicity for Dayton, the town at the centre of the controversy.

The state government in Tennessee passed a law that is often referred to as the Butler Act or Butler's Law after John Washington Butler, the state legislator who proposed it. This banned the teaching of evolution in schools but also insisted that teachers use a particular textbook containing a chapter that covered the theory of evolution. Thus, anybody using the set textbook would effectively be breaking state law. To test this law, the American Civil Liberties Union said that it would defend anyone accused of teaching evolution theory in school. With his own agreement, John Thomas Scopes, a high school teacher, was then arrested so that the issue could be brought to court. Clarence Darrow, a famous labour lawyer, agreed to lead the defence. William Jennings Bryan, a well-known late 19th-century reformer and lawyer, was asked by the World Christian Fundamentals Association to lead the prosecution.

Discussion point

Interestingly, in her book, Christine Garwood makes the observation that although he was a defender of "fundamental Christian faith and creationist teaching in schools", he was criticized for not being a "true fundamentalist if he believed the earth was a globe" by Flat Earth believers. However, she also points out that major creationist organizations – such as the Bible-Science Association, the Christian Research Society, and the Institute for Creation Research – all "recoil from association with Flat-Earth believers".

Bryan wanted to counteract the idea that Darwin's evolution implied a death struggle, the "survival of the fittest". He pointed out that during the First World War, many German intellectuals and military leaders invoked Darwinism as the justification for war and future domination, whilst in England and America it was also used as a justification for industrial exploitation through natural selection. This also came to be known as "social Darwinism", and was a fundamentally incorrect interpretation and misuse of Darwin's theory, as it implies that natural selection only means a struggle for survival rather than the possibility of mutual aid.

As Gould highlights in *Bully for Brontosaurus*:

1. Evolution means "only that organisms are united by ties of genealogical decent", which "says nothing about the mechanisms of evolutionary change".
2. Darwin's theory of natural selection "is an abstract argument about a metaphorical 'struggle' to leave more offspring in subsequent generations, not a statement about murder and mayhem".
3. And that "whatever Darwinism represents on the playing fields of Nature, it implies nothing about moral conduct".

Originally, Darrow was going to base the defence's argument on the violation of an individual's rights, but changed it to show that there was no conflict between evolution and the creation account in the Bible. To do so, he called eight expert witnesses to prove his argument, only one of which was allowed to testify to the judge. The others were permitted to submit written statements.

The prosecution argued that the scientific testimony was neither competent nor proper, as the case was simply whether Scopes had or had not taught evolution at his school.

After eight days, the trial ended with a guilty verdict and Scopes was ordered to pay a $100 fine. The verdict later went to appeal and the conviction was eventually set aside because of a legal technicality.

Tennessee repealed the Butler Act in 1967, and in 1968, the Supreme Court of the United Sates invalidated a similar law in Arkansas, which required that evolution and creation science should be taught equally. This, the Balanced Treatment for Creation-Science and Evolution-Science in Public School Instruction Act, did not require teachers to teach either evolution or creation science, but that when evolution was taught, creation-science had to be taught as well.

The Supreme Court repealed the law on three counts, known as the Lemon Test:

1. The government's action must have a legitimate secular purpose.
2. The government's action must not have a primary effect of either advancing or inhibiting religion.
3. The government's action must not result in an "excessive entanglement" of the government and religion.

Since the 1960s, there have been at least 16 other legal cases in America that have been used to test areas of the law that cover the teaching of evolution in schools. These have covered issues such as attaching disclaimer stickers to textbooks, reading out disclaimer notices before lessons, and clarifying whether a teacher, lecturer, or professor has the right to teach particular materials due to his/her religious beliefs. Each time the legal "test" has been based on the above Lemon Test, and it appears that on each occasion the teaching of evolution as a valid theory has won through.

7.3.4 The nature of life and science, and Evolution versus Creationism

The debate between evolution and creationism centres on the philosophies of both, and whether it is wrong to take science out of context.

As we saw in Chapter 4, when Kirwin attacked Hutton's views in 1799 concerning his ideas of the Earth, he took them out of context and implied that it meant that the Earth was eternal. In fact, Hutton's entire argument was based on his attempt to develop a cyclic theory to match that of Newton's planets revolving around the Sun. His theory said that we cannot learn anything about the formation and the end from Nature's present laws. He thought that speculations on the origin of the Earth – when not based on observable facts – did not qualify as "proper science". Similar arguments were used in the American Supreme Court in 1968, to stop the equal time laws.

Having looked at the nature of stratigraphy in Chapter 2, construction of the geological time scale in Chapter 3, Neptunism and Plutonism in

Chapter 4, Uniformitarianism and Catastrophism in Chapter 5, and Evolution in Chapter 6, it is important to include examples of continuing debate between evolutionism and creationism with regard to geology and the nature of science.

According to Gould, evolution does not try to study the origins of life – which he says falls to chemistry and physics – it studies the pathways and mechanisms of organic change. As he puts it, "we have oodles to learn about how evolution happened, but we have adequate proof that living forms are connected by bonds of genealogical descent". Following Hutton's wisdom, he says, "we do not search for unattainable ultimates".

As Lucas says in *Genesis Today: Genesis and the Questions of Science*, "scientific questions can always be expressed as 'how?' questions". He adds that "science can never answer questions of meaning, the 'why?' questions. Scientific attempts to answer them always end up as 'how' answers". In other words, science and religion ask different questions that result in different answers.

An alternative way to describe how science works is provided by Walter Alvarez when he informs us that scientists are "engaged in a conversation with Nature. We ask questions – like 'Where is the crater?', by making observations or performing experiments". Nature then provides the answers as long as we have the ability to find, observe, and interpret them. He warns however that "a young scientist, just starting out, cannot imagine how hard it is to understand the real meaning of Nature's answers, or how many ways there are to make mistakes and get fooled". As for his own experiences, in chapter 5 of his book, he described the problems encountered with trying to track down the K-T boundary impact crater. In spite of the evidence, he and his colleagues frequently drew the wrong conclusions that led them in the wrong direction, only to discover their mistake some time later.

As Palmer explains, although the history of establishing the geological succession had largely focused on identifying, dividing, and grouping strata during the early to middle 19th century, the:

> "… brotherhood of the hammer" were highly territorial and ambitious as they hustled and bustled amongst the rocks and argued about how to divide up Earth Time. They hoped to promote themselves and secure lasting personal reputations by establishing their "systems" as nationally and internationally recognized divisions of Earth Time.

Geologist's usually focus on facts, i.e. rock descriptions and relationships between different rock units together with the fossils they contained, rather than theories or explanations. One of the problems they have always had to contend with is the variability of apparently "rock solid facts". Palmer sums up the situation well when he says that "disputes were and still are the name of the game and ranged from minor spats to long-running, bitter, and highly personalized feuds between the protagonists. Whoever said that science was impersonal?"

Palmer examines the arguments between geologists and biologists over the discovery of life in the Precambrian by geologists such as Walcott. Following these discoveries, biologists such as Sir Albert Charles Seward (1863–1941), a botany professor at the University of Cambridge, were "dismissive of the evidence for Precambrian life", with such phrases as "we can hardly expect to find in Precambrian rocks any actual proof of the existence of bacteria". Palmer goes on to say, "but in science, as in many other aspects of life, never say never. Seward turned out to be spectacularly wrong". For example, Prothero includes descriptions of microscopic fossils found in the Warrawoona Group in Western Australia and the Fig Tree Group in South Africa. The former include cyanobacteria and stromatolites that are around 3.5 billions years old, whilst the latter contain stromatolites dated as 3.4 billion years old. Both exhibit structures that are "virtually indistinguishable from their modern counterparts".

Prothero also adds a valuable observation concerning the views of Creationists on the "Cambrian Explosion" of life. Fossil evidence now indicates that it was more of a "slow fuse" than an explosion, but:

> Creationists love to quote a variety of legitimate scientists about the "mystery" of the Cambrian explosion, although most of their quotes are grossly out of date, and many are out of context and say just the exact opposite when the full quote is read carefully.

In fact, this may be true of many Creationist arguments.

Discussion point

It has been said that there should be no conflict between science and religion, as one looks at things from a "how" point of view and the other from a "why" point of view, but is this a valid argument or is it simply a way of avoiding the debate?

Should the government/state intervene in what should be taught in schools, colleges, and universities?

In chapter 14 of his book *The New Creation: Building Scientific Theories on a Biblical Foundation*, Garner reviews a modified version of the Ecological

Zonation Theory, originally proposed by Harold Clark in 1946. In this, life was separated out into adjacent ecological "provinces" that when "reconstructed are very different from those found on the surface of the Earth today". These include an interpretation in which Precambrian life existed in a "stromatolites reef community" in a hydrothermal environment around the margins of pre-Flood continents. Garner also proposed the existence of a "floating forest community" that presumably accounts for the formation of the Carboniferous Coal Measures. In these explanations, he uses specific examples to highlight irregularities that conform to creationist models. He also uses sweeping generalities that avoid or are devoid of specific content. He even re-introduces the idea of changes in ocean salinity (Chapter 1) as evidence for a young Earth.

In *Geology in the Bible*, Billy Caldwell outlines a number of geological phenomenon. He then uses selected, sometimes misleading examples, or omits important information, to conclude a young age for the Earth. He also makes statements such as:

> The concept and belief in uniformitarianism has also led geological history astray, and evolution … is the great lie of this century. Most of the colleges teach a godless creation of the Earth. The instructors who teach this have much head knowledge and little wisdom.

He informs his readers that "geologists have studied the rocks for many years and have pieced together a geological history of the Earth from this information. This study has been totally influenced by humanistic ideas". He concludes that the age of the Earth was determined by evolutionists, even though it is quite clear that the reverse was true.

Discussion point

As noted in Chapter 3, the geological time scale was established for the most part by geologists who believed in the existence of successive creations, that is, not evolutionists (because the concept of evolution postdated their work). In other words, it was not put together to act as a "proof" for evolution – as has been suggested by a number of creationist authors – but was a serious and successful attempt to put the geological past into its correct order, and into a relative time scale.

Jonathan Wells has written an interesting book titled *The Politically Incorrect Guide to Darwinism and Intelligent Design*. This largely comprises a rant against Darwinism and science in general, with little explanation of

Intelligent Design (ID). This could be compared to Mark Isaak's brief but clear explanation in *The Counter-Creationism Handbook.*

Eugenie Scott's chapter "Creation Science Lite: 'Intelligent Design' as the New Anti-Evolutionism", in Petto and Godfrey's book, *Scientists Confront Creationism: Intelligent Design and Beyond*, makes an interesting observation concerning the definition and use of the term Darwinism:

> This fixation on "Darwinism/Darwinist" in ID literature is puzzling to scientists, who after all, do not refer to physicists as Kelvinists or geologists as Lyellists. In evolutionary biology, "Darwinism" usually refers to the general idea of evolution by natural selection; it may specifically refer to the ideas held by Darwin in the 19th century. Usually the term is not used for modern evolutionary theory, which, because it goes well beyond Darwin to include subsequent discoveries and understandings, is more frequently referred to as "neo-Darwinism," or just "evolutionary theory". … In ID literature, however, "Darwinism" can mean evolution itself, natural selection, Darwin's ideas, or neo-Darwinism, but most commonly it refers to materialist ideology inspired by "Godless evolution".

Interestingly, Wells almost invariably refers back to Darwin's original ideas when arguing against evolution, ignoring those within neo-Darwinism, as if there have been no changes.

With regard to extinctions, evolution, and Intelligent Design, Jerry Coyne comments:

> It's important to realize, though, that there's a real difference in what you expect to see if organisms were consciously designed rather than they evolved by natural selection. Natural selection is not a master engineer, but a tinkerer. It doesn't produce the absolute perfection achievable by a designer starting from scratch, but merely the best it can do with what it has to work with.

Turning to extinctions, he observes that:

> This, by the way, poses an enormous problem for theories of intelligent design. It doesn't seem intelligent to design millions of species that are destined to go extinct, and then replace them with other, similar species, most of which will also vanish …

Discussion point

Is the ongoing debate between evolution versus creation really about science, or the positions of science and religion?

The International Society for Science and Religion was established in 2002. Its purpose was to promote education through the support of inter-disciplinary learning and research in the fields of science and religion in an international and multi-faith context. The society includes the following as part of its statement on ID:

> We believe that intelligent design is neither sound science nor good theology. Although the boundaries of science are open to change, allowing supernatural explanations to count as science undercuts the very purpose of science, which is to explain the workings of Nature without recourse to religious language. Attributing complexity to the interruption of natural law by a divine designer is, as some critics have claimed, a science stopper.

Further reading

Whilst much of the ID argument is focused on biological and biochemical examples, they also delve into geology, so we cannot afford to ignore them. The book *For the Rock Record: Geologists on Intelligent Design*, edited by Jill Schneiderman and Warren Allmon, comprises a series of essays, written by a number of geologists that consider ID arguments with regard to a range of geological topics, each of which makes interesting reading.

It is also worth reading chapters 7, 8 and 9 of Eugene Scott's book, *Evolution versus Creationism: An Introduction*, which presents a number of arguments, together with legal and educational issues involved in the "battle" between creationists and evolutionists. Although she presents this as primarily an American issue, similar situations are being raised here in Britain. Mark Isaak's *The Counter-Creationism Handbook* also provides well-written, comprehensive answers to numerous Creationist claims. *Monkey Trials & Gorilla Sermons* by Peter Bowler also contains an extremely good overview of the origins of the conflict between Darwinism and Creationism, and the rise of Intelligent Design. Neil Shubin's *Your Inner Fish* details research that rebuffs many ID claims.

Finally, Donald Prothero's *Evolution: What the Fossils Say and Why It Matters* makes interesting reading, particularly his discussion in chapter 2 (Science and Creationism), which covers relationships between Creationism, Scientific-Creationism, and Intelligent Design. He concludes that they are basically the same, with the same ideals and goals but each wrapped in a different cloak. His final chapter makes interesting but chilling reading, particularly with regard to people who say that even discussing modern creationism gives it too much credence.

7.4 Lagerstatten

Finally, before we leave evolution and creation, it is worth mentioning sites where the range and diversity of life forms and the detail of preservation is exceptional. These are known as *Lagerstatten*, meaning fossil deposit places, in German. It can also be translated as "fossil-bonanzas" or "rock bodies usually rich in palaeontological information". These give us "snap-shots" of the diversity of life in the past. In each case they show that life was far more diverse than the normal fossil record would tend to indicate (see Palmer's comments included in the introduction to this chapter) and, because of the exceptional level of preservation, the fossils also show us details that are usually missing, which allow us to gain a clearer understanding of how they work and how they have developed. Even as I am writing this, there are more reports that fossil evidence has been found indicating that Velociraptors – the demons of Jurassic Park – had feathers. Although the hunting habits still remain, when you consider that generally they were the size of a large chicken or turkey, the addition of feathers changes our perception of them.

Lagerstatten are therefore very important sequences that help to answer important questions. As Euan Clarkson quotes in his book, *Invertebrate Palaeontology and Evolution*, "only a fraction of the myriad creatures that have lived on the Earth have left behind traces of their existence, and only specific parts of those organisms have been preserved". He adds, "normally we expect to see no more than a narrow band of 'preservable' organisms from an originally much broader biotic spectrum". This means that our perception of the course of evolution is primarily based on the hard parts that have been preserved from a small number of animals.

Add to this the comments of Philip Manning that:

> The depths of geological time and the breadth of life that has evolved through it provides one of the most enthralling stories the planet has to offer. We often dwell on human history, but that is no more than a single breath of our planet's long life. While the fossil record is by no means complete, the occasional glimpse of "wonderful things" allows insight to this ultimate story of life. Rather than complete books of the "great works of life on Earth", many pages and chapters, if not volumes, are yet to be discovered.

Lagerstatten give us glimpses of the extraordinary diversity of life that has existed on Earth that are otherwise missing. These enable geologists and palaeontologists to make huge advances in knowledge and interpretation concerning the potential diversity of life in the past.

Table 7.1 shows many of the Lagerstatten sites around the world. These provide an indication of the number and distribution of the "snapshots" that exist. As you can see, they cover most geological time periods, but their limited spatial distribution tends to indicate that there could be more sites yet to be discovered.

Discussion point

Do you think that the ever-increasing number of fossils being discovered around the world will provide evidence for evolution of creation?

Table 7.1 Lagerstatten sites around the world (some of the dates are only approximate values)

Geological period	Site	Location	Age (Ma)
Pleistocene	La Brea Tar Pits	California, USA	20,000 yrs
Miocene	Clarkia fossil beds	Idaho, USA	17–20
Oligocene-Miocene	Dominican amber	Dominican Republic	10–30
	Riversleigh	Queensland, Australia	15–25
Eocene	London Clay	England	48–54
	Green River Formation	Western USA	48
	Princetown Chert	Canada	49
	Monte Bolca	Italy	49–52
	Grube Messel Shale	Frankfurt, Germany	49
	Messel Oil Shale	Hessen, Germany	49
Cretaceous	Pierre Shale	North Dakota, USA	80
	Auca Mahuevo	Patagonia, Argentina	80
	Santana Formation	Brazil	92–108
	Crato Formation	Northeast Brazil	92–108
	Hajoula Limestone	Lebanon	93–97
	Tlayua	Mexico	100
	Xiagou Formation	Gansu, China	105
	Jehol Group	China	120–133
	Yixian Formation	Liaoning, China	121–125
	Las Hoyas	Spain	121–127
	Sierra de Montsec	Spain	130–135

Table 7.1 *Continued*

Geological period	Site	Location	Age (Ma)
Jurassic	Purbeck Beds	England	137
	Morrison Formation	Wyoming, USA	147–156
	Solnhofen Limestone	Bavaria, Germany	155
	Christian Malford	England	158
	La Voulte sur Rhone	France	158
	Stonesfield Slates	England	163
	Holzmaden	Wurttemberg, Germany	185
	Posidonia Shale	Germany	185
Triassic	Karatau	Kazakhstan	144–213
	Ghost Ranch	New Mexico, USA	205–286
	Gres a Voltzia	France	246
Permian	Wellington Shale	Kansas, USA	285
Carboniferous	Karoo System	Zimbabwe, Southern Africa	208–286
	Hamilton Quarry	Kansas, USA	295
	Mazon Creek	Illinois, USA	300
	Bear Gulch Limestone	Montana USA	318
	Loch Humphrey Burn	Scotland	330
	Scottish "Shrimp Beds"	Scotland	345
	East Kirkton	Scotland	345
Devonian	Gogo Formation	Australia	350
	Cleveland Shale	Ohio, USA	354–417
	Canowindra	New South Wales, Australia	360
	Escuminac Bay	Canada	370
	Gilboa	New York, USA	380
	Hunsruck Slate	Rhineland, Germany	390
	Rhynie chert	Scotland	396
Silurian	Fiddler's Green Formation	New York, USA	410
	Wenlock Series	England	423–428
	Waukesha	Wisconsin, USA	425
	Lesmahagow	Scotland	425
Ordovician	Soom Shale	South Africa	435

Table 7.1 *Continued*

Geological period	Site	Location	Age (Ma)
Cambrian	Oland Orsten	Sweden	500
	Kinnekulle Orsten	Sweden	500
	Andrarum Limestone	Sweden	500
	Wheeler Shale, House Range	Utah, USA	500–540
	Burgess Shale	British Columbia, Canada	505
	Kaili Formation	Guizhou Province, China	506–513
	Chenjang	Yannan, China	515–520
	Sirius Passet	Greenland	518–520
	Emu Bay shale	South Australia	522
	Maotianshan Shales	Yunnan Province, China	523
Precambrian	Doushantuo Formation	Guizhou Province, China	570
	Ediacara Hills	South, Australia	565–595

8
Continental Drift and Plate Tectonics

8.1 Introduction

Plate Tectonics is crucial to most modern geological thinking and therefore an understanding of how the theory has developed is important. Like many of the other subjects covered in this book, its development is contorted, spread over a considerable period of time, and involves a great deal of lively debate. As with most of the other subjects, this chapter is not designed to present the details of the Continental Drift or Plate Tectonic theories – you can find these in most geological or physical geography textbooks. There are also plenty of other books available that cover the subject in a variety of ways and depths. The following sections are designed to show you how different people were involved in developing both theories and how others accepted or rejected their ideas for a variety of reasons.

Discussion point

As you read through the following sections, it should become apparent that – as with many other areas of geology – lines of evidence were accepted and rejected by different people at different times. This is interesting, as we tend to think that most of the controversial debates happened in the 19th century, whereas Plate Tectonics is a relatively new theory and its predecessor, Continental Drift, is not that much older. You will see that even in the relatively recent past, geologists were not immune from being selective about what they agreed or disagreed with.

One of the reasons for covering these theories in the way that they are presented here is because it involves a fundamental scientific approach, which says that all theories have equal merit until sufficient evidence is found to support one or more above the others. This seems to be a very good way of approaching science – especially when you think back to other examples in this book, when particular personalities or ideas have been pursued, only to discover later that they were at best incomplete answers or at worst completely wrong.

Time Matters: Geology's Legacy to Scientific Thought, 1st edition. By Michael Leddra.
Published 2010 by Blackwell Publishing Ltd.

Let us begin with the simple questions, how good is Plate Tectonic theory? In addition, how long will it last in its present form? At the moment, it is central to all geological ideas. There is a huge variety of information that fits the pattern, and for the first time it appears that we have a truly global model, but past experience indicates that sooner or later it will change. In reality, we do not know what scientific revolutions are "around the corner". It could well be that at some time in the future we will discover some piece of information that may completely overturn Plate Tectonic theory as we know it today, or at least lead to the development of a significantly modified version of it that may even require a new name. The fact is that we do not have all the answers – there are still bits of the jigsaw puzzle missing. At present, it is the best model we have.

As Philip Kearey and Frederick Vine record in their book, *Global Tectonics*, "ever since man first charted the coastlines of the continents around the Atlantic Ocean in the 16th century, he has been intrigued by the similarity of the coastlines of the Americas and of Europe and Africa". They point out that the first person to "note the similarity and suggest an ancient separation" was Abraham Ortelius in 1596. They also write that in 1620 Sir Francis Bacon spoke about the similarities in shape between the coasts of South America and Africa, and in 1666 Francois Placet suggested that Europe and Africa may have been separated from the Americas by either a floating island or the destruction by Noah's Flood of an intervening landmass – which could have been "Atlantis". Theodore Lilienthal – a Professor of Theology from Konigsberg, Germany in 1756 – considered Noah's Flood to be the cause of a split between America and Africa. In both 1801 and 1845, Alexander Humbolt thought that the Atlantic had been formed by the same flood and that the ocean was "a valley scooped out by the sea". Finally, Kearey and Vine record that in 1858 it was Antonio Snider-Pellegrini (1802–1885), a French geographer, who proposed the idea that the continents may have drifted apart due to multiple catastrophes, in which Noah's Flood was only the last of many. He also produced a map, which showed what he thought the world would have looked like when the continents were joined together: this map is remarkably similar to the maps produced in the last 30 years.

The following is largely based on *The Rejection of Continental Drift, Theory and Method in American Earth Science* by Naomi Oreskes. It is well worth reading the entire book, as it contains a very detailed and well-referenced account of the different approaches to scientific thinking in America and Europe. It is interesting to see how – even relatively recently – scientists on different continents or even in different countries have had significantly different approaches to the same topics. Generally, this appears to be based on the way, historically, in which science has been studied in each country or continent.

8.2 Mountain building

One of the significant areas of interest in the early 1900s was the origin of mountains. A major focus of ideas centred on their formation through compression due to the gradual thermal contraction of the Earth; in other words, as the Earth cooled it shrank and crumpled. There were two important players in this area: Eduard Suess and James Dana.

Eduard Suess (1831–1914) (Fig. 8.1) was a field geologist from Vienna, who spent his life looking at the structure of the Alps. His ideas were published in English, between 1885 and 1901, in four volumes of *The Face of the Earth*. He thought that the Earth originally had a continuous continental crust, which gradually broke up and wrinkled as the Earth cooled and contracted. Contraction also caused some areas of the crust to collapse, forming the oceans. As the Earth continued to contract, the uplifted continents became unstable and collapsed to form new oceans,

Fig. 8.1 **Eduard Suess**

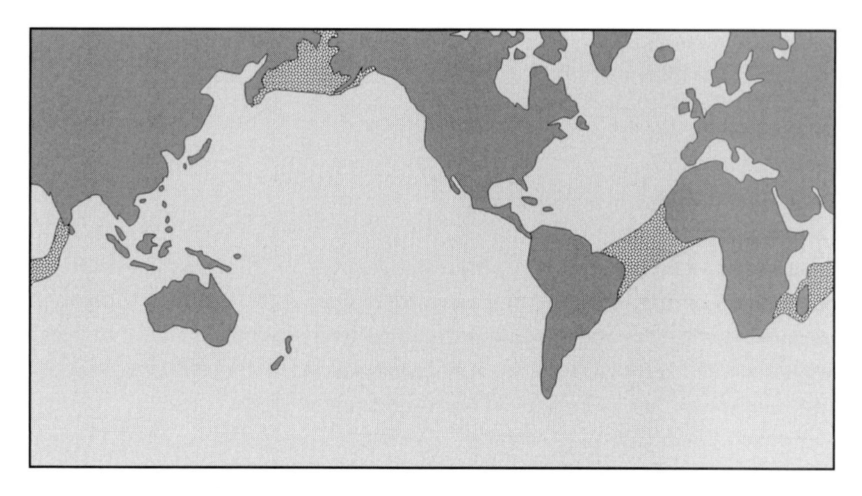

Fig. 8.2 **The location of proposed land bridges (dotted areas) between each of the major continents**

and the old oceans buckled up to form new continents. He termed his original supercontinent (which covered the Earth) "Gondwanaland" after the Gondwana System, a sequence of rocks found in the Gondwana region of Central India. He considered the sunken areas to be submerged "land bridges". Suess also thought that his idea, known as the "Contraction Hypothesis", would account for the formation of mountains on every continent and igneous activity associated with the edges of the sunken blocks (Fig. 8.2).

The idea of thermal contraction (cooling from the centre of the Earth) had been around for a long time: it was based on the theory that the Earth had formed from the condensation of hot gases and was still cooling down (the Kant–Laplace Nebular hypothesis of 1796). De La Beche (1796–1855) used this idea to explain sudden, periodic geological movements in opposition to Lyell's idea of uniformitarianism with its slow, steady changes. Elie De Beaumont, Jean Baptiste Armand Louis Leonce (1798–1874) – a French geologist – used the same idea to account for the fact that the orientation of folds and mineral veins in mountain chains did not have a random distribution or orientation, as both were related to the age of their formation and the global forces involved in the formation of the mountains.

Before we leave Suess, it is also worth mentioning that Simon Winchester in his book, *A Crack in the Edge of the World*, points out that it was Suess who first recognized the three-rock sequence that comprises the Oceanic Crust – namely a gabbro/lava sandwiched between serpentinite (below) and deep-sea sediments, including radiolarian chert (above) that together are known as an "ophiolite". The name is derived from the Greek words *ohpis*, meaning "snake" and *lithos* meaning "rock", as serpentinite (which is also known as soapstone) often has a very distinctive green-blue or blue-red colour, which looks a bit like the skin of a snake.

In America, James Dwight Dana (1813–1895), a geologist, mineralogist, and naturalist, proposed a different version of thermal contraction, known as the "Permanence Theory", in which mountains and oceans were permanent features. This was based on his knowledge of mineralogy and astronomy: he thought that the moon's craters and mountains were both formed of similar materials and by similar processes, but later in the moon's history the craters cooled and sank. He noted that most mountain chains on Earth form along the edges of continents, and theorized that they were formed by "lateral pressures" caused by the subsidence of the adjacent oceans. He also thought that the highest mountains would be adjacent to the deepest oceans and that the ocean floors would be composed of a different material to that of the continents. By the start of the 20th century, in America, this was considered to be a fact.

Another proposal was made by James Hall (1811–1898), a geologist and palaeontologist, was based on the presence of thick sedimentary sequences in the Appalachian Mountains of America. His idea also took into account the formation of areas of large-scale subsidence and deposition adjacent to the edges of continents, which Dana had called "Geosynclines" (Fig. 8.3). The formation of geosynclinal basins led to further depression along the margins of the surrounding continents, which allowed the sediments that had been deposited in the basins to be heated and compressed. Eventually these heated and compressed sediments were uplifted to form mountains. It was thought that the apparent periodicity of mountain formation found in the geological record was therefore due to the time it took for sediments to accumulate in geosynclines. Dana also proposed that sediment build-up in geosynclines did not cause them to subside, but that subsidence in the upper part of the mantle allowed a build-up of thick sequences of sediments.

As Oreskes puts it, "Dana's was a unifying theory, bringing together the best of American fieldwork with the long-standing European tradition of interpretation based on the premise of secular cooling". By the end of the 19th century, there were two different views based on the same starting point – "in the European view, the Earth was in a state of continual flux with complete interchangability of its parts", whereas "in the American view, the basic outlines of the Earth had been set at the beginning of geological time and had not changed fundamentally since then". The American view was based on mineralogy and the difference in terrestrial and oceanic materials. The European view was based on biogeographical patterns,

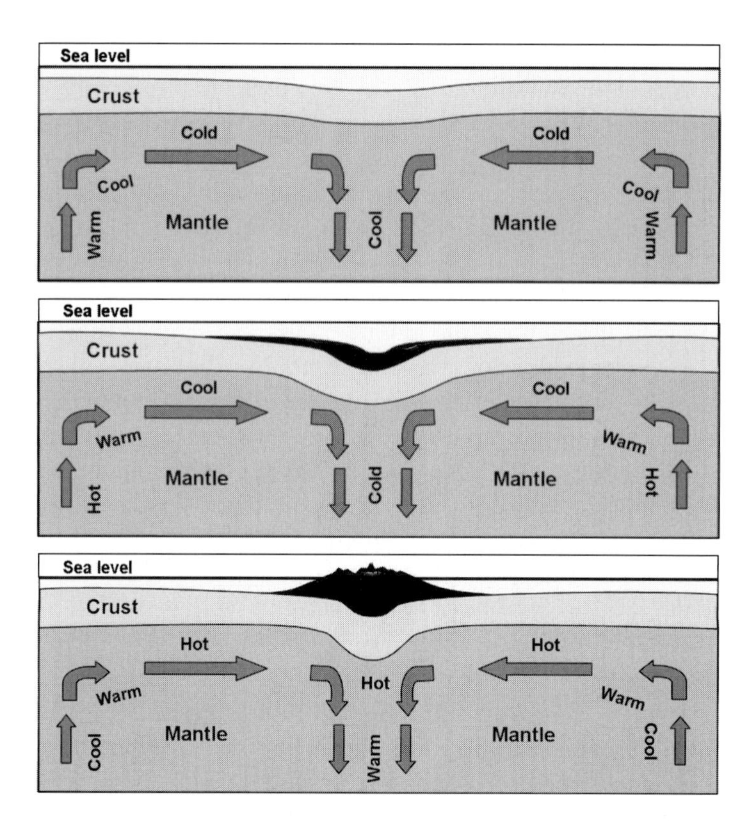

Fig. 8.3 **A diagrammatic representation of the formation of a geosyncline. Initially the crust sinks due to the effects of cold currents in the upper mantle (top diagram). Large-scale sedimentation fills subsiding basin (centre diagram), as the temperature of mantle circulation then warms up. Finally, the sediments are compressed to form fold mountains (lower diagram)**

stratigraphy, and the diverse pattern of folding in mountain belts. Oreskes also points out that in Britain, "neither theory was entirely accepted" – in other words, we were neither "Americans" nor "Europeans".

8.3 Isostasy

Isostasy is a condition that essentially relates Archimedes Principle to the Earth, in which the lithosphere – the Earth's crust – "floats on" the Earth's upper mantle, rather like an iceberg (Figs 8.6 and 8.7). Changes in the thickness or density of the lithosphere lead to changes in the degree to which it "sinks" into the upper mantle (termed the asthenosphere).

Fig. 8.4 **Clarence Dutton**

Reverend Osmond Fisher (1817–1914), a geologist at the University of Cambridge, first proposed this idea whilst he was trying to formulate equations to "prove" or at least quantify the Contraction Hypothesis. His work actually proved that contraction could not account for differences in observed elevation of mountains. Fisher also studied the geomorphology of Norfolk, the stratigraphy and fossils of Dorset, and published *The Physics of the Earth* in 1881, in which he described the mechanisms of convectional currents in the Earth's interior.

The term "Isostasy" was proposed by the American geologist Clarence Dutton (1841–1912) (Fig. 8.4), one of the founders of seismology (the study of earthquakes). He thought that the Earth had originally been liquid which, as it cooled, formed a solid core and crust with a fluid layer between them. This proposition was in direct opposition to the ideas of Kelvin, who insisted that the Earth was solid and rigid. Dutton proposed that the fluid layer beneath the crust allowed for the formation of mountains through horizontal movement and compression (stress). Fisher also thought that the Oceanic Crust had a different density to the Terrestrial or Continental Crust.

Dutton developed the idea of Isostasy to replace the Contraction Hypothesis, as the method for explaining the formation of large-scale features. He thought that as rocks were eroded, the materials they produced were deposited as sediments in the oceans. As the layers of sediments built up, their weight depressed the crust, causing it to heat up, expand, and "flow" to produce igneous intrusions, which then led to uplift in the eroded areas. When the balance between the Continental Crust and its substrate are in balance, this is known as "Isostatic Equilibrium".

How does isostatic equilibrium operate? Dutton and Fisher thought that erosion reduced pressure (and weight) and sedimentation increased it. A well-known example of isostatic equilibrium is the gradual tilting of the UK in response to the removal of the weight of ice that covered Scotland, Wales, and northern and central England during the last glaciation. In response, each of these areas are rising whilst southern England is tilting downwards, rather like the rocking of a seesaw until the movement comes back to equilibrium or stability.

Suess' Contraction Hypothesis ran into a number of problems by the start of the 20th century, after it had been recognized in the 1840s that the Alps were formed by horizontal rather than vertical movements. This resulted in a major problem, as the primary movements in the Contraction Hypothesis were vertical. George Everest (1790–1866) – Surveyor-General for India and after whom Mount Everest was named – and John Henry Pratt (1809–1871), the British clergyman and mathematician who devised the theory of crustal balance on which the theory of Isostasy is based – recorded differences in gravitational attraction during geodetic surveys in India. These surveys indicated that the Himalayas had a smaller mass than predicted and that coastal areas appeared to have a higher density than expected. Consequently, two different models of Isostasy were developed – one by George Airy, which required a "fluid substrate" and the other by John Henry Pratt, which did not.

George Biddell Airy (1801–1892) (Fig. 8.5), the English Astronomer Royal, suggested that the difference in density in the Himalayas was due to a thickening of a low-density crust (known as the lithosphere) into a higher-density substrate (the asthenosphere) – i.e. differential thickness, rather like an iceberg, where the crust is thicker under mountainous areas and thinner under planes (Fig. 8.6).

On the other hand, John Henry Pratt proposed that these differences were due to variations in density between the rocks of which mountains, planes, and oceans were composed. He thought that this meant that the crust had a constant thickness (Fig. 8.7). These two hypotheses were known as the Airy's Roots of Mountains Hypothesis and Pratt's Uniform Depth of Compensation Hypothesis, where the depth of compensation is the level within the Earth at which the masses (weights) of all the overlying features are the same.

Discussion point

It should be born in mind that American geologists had a different approach to looking at scientific theory, which determined how they assessed each idea. A good example of this is that of Thomas Chrowder Chamberlin (1834–1928), a very prominent American geologist, who thought that science as well as society should be able to operate as a democracy and therefore all theories should be treated equally. Oreskes quotes something he wrote in the *Journal of Science* in 1890, to substantiate this notion:

> The general application of this method to the affairs of social and civic life would go far to remove those misunderstandings, misjudgements, and misinterpretations which constitute so pervasive an evil in our social and political atmosphere, the sources of immeasurable suffering ... I believe that one of the greatest moral reforms that lies immediately before us consists in the general introduction into social and civil life of that habit of mental procedure which is known as the method of multiple working hypotheses.

He thought this should be the approach used in science as well; an approach, which effectively laid down the rules for American thinking. This means that science should progress through consensus and the view of the majority.

Chamberlin developed his *Planetesimal Theory*, which proposed that the Earth formed by the gradual build-up of small objects. He then used this theory to propose that Kelvin's estimate for the age of the Earth (i.e. 100 million years old) was far too small.

John Fillmore Hayford (1868–1925) – a geodesist who published *The Figure of the Earth*, and who is also well-known for the Hayford Spheroid – looked at 500 geodetic survey stations throughout the USA and found that the principle of Isostasy could account for the systematic differences between the values of gravity recorded and the theoretical values for each of the sites. Using this data and Pratt's model of Isostasy, he managed to "establish Isostasy as a fact". He also decided that it proved that the continents and oceans could not be interchangeable because, as their densities were different, they had to be permanent features. This fitted with Pratt's model and Dana's version of a Contracting Earth.

The discovery of radioactivity in 1903 by Marie Curie effectively destroyed the basis of a Contracting Earth, through the discovery that heat was produced by radioactive decay, so now geologists knew that they were finally dealing with an Earth that did not have to cool down. Although this

Fig. 8.5 **George Biddell Airy**

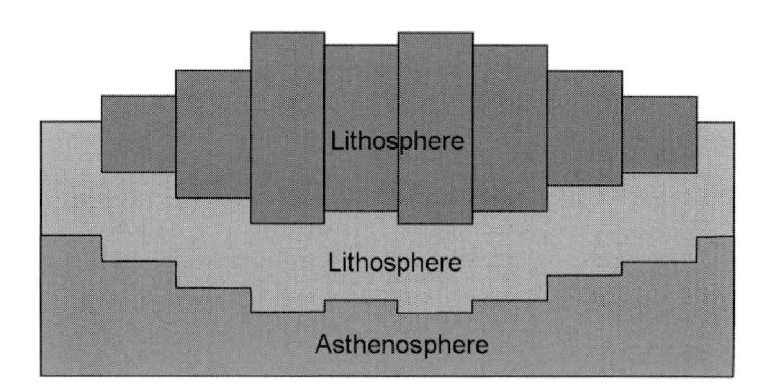

Fig. 8.6 **A diagrammatic representation of Airy's model of Isostasy**

solved one problem, they were faced with others: if the Earth was not shrinking, how could they account for fossil and stratigraphic evidence, which indicated that sea levels were different in the past compared to the present?

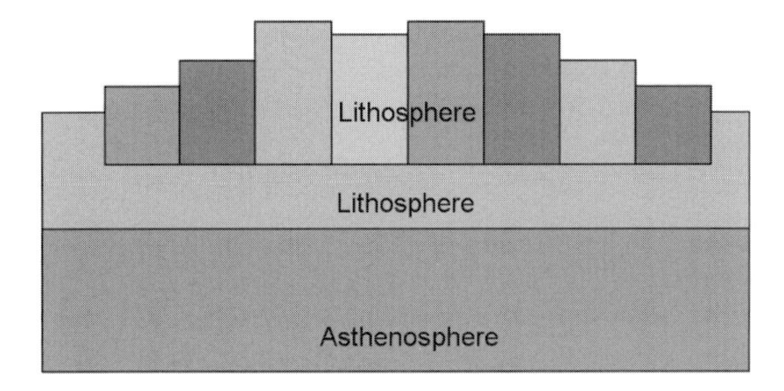

Fig. 8.7 **A diagrammatic representation of the Pratt model of Isostasy. The different shades in the continental lithosphere represent different densities**

Fig. 8.8 **Alfred Wegener**

8.4 Continental Drift

Philip Kearey and Frederick Vine record that the American physicist Frank Bursey Taylor suggested the first uniformitarian view of Continental Drift in 1910, to explain the origins of fold mountains. However, it was Alfred Wegener (1880–1930) (Fig. 8.8), a German scientist and meteorologist who is usually credited with the concept of Continental Drift and who also proposed his "Displacement Hypothesis" in 1912. Over the next 10 years, he developed this into a theory that he outlined in a book called *The Origin of Continents and Oceans*, first published in 1915. He proposed that the continents slowly drifted through the ocean basins and occasionally collided with each other. To start with, only a few people in Britain and Europe took any notice of the idea. In America, his ideas were completely rejected – but why? The basic elements of Wegener's theory were that:

1. Continents consisted of less dense materials, which caused them to "float" in hydrostatic equilibrium within a denser oceanic substratum.
2. They move because, over geological time, the substratum behaves like a highly viscous fluid.
3. Originally, there was a thin continuous Continental Crust, which gradually broke up and thickened because of crumpling of the moving pieces.
4. During the Mesozoic Era, the major continents came together to form a supercontinent in the Southern Hemisphere, known as "Gondwanaland".

Many of these were effectively bringing Isostasy and Suess' theories together. However, Wegener thought that Isostasy could not exist in a shrinking Earth and that fossil and stratigraphic evidence indicated large changes in the position of the oceans, which did not fit with Suess' ideas. Many geologists in Europe took a different approach – they accepted the fossil and stratigraphic evidence that Wegener presented, but felt that the existence of land bridges rather than Isostasy provided the best solution. They therefore tended to reject the idea of Isostasy, which was primarily an American theory. Wegener was effectively trying to bring two different views from two different continents together, i.e. that continents were permanent but that they tend to move horizontally rather than vertically, and he was also trying to combine both of Pratt's and Airy's models of Isostasy.

So what was Wegener's evidence? Oreskes – like many other authors – points out that as far as Wegener was concerned, the palaeontological evidence was the most important:

> ... perhaps because the data were already established and accepted in a different context. In the 1850s the British zoologist Philip Sclater had noticed that the island of Madagascar possessed almost none of the common African animals such as monkeys, giraffes and lions but hosted numerous species of lemurs, an animal common to India.

She also reports that "By the 1920s former connections had been postulated between Australia and India, Africa and Brazil, Madagascar and India, and Europe and North America". These continents also contain common fossil species, many of which would have been unable to move from one continent to another across water. Oreskes also adds that fossil evidence indicated that if "the continents had not moved, the required connections would have been extremely long and crossed several climatic zones".

The jigsaw image of the continents – or more correctly the edges of their continental shelves – had also been recognized for a long time, as had the extraordinary similarity between stratigraphic sequences and structural elements (such as fold mountain belts), which matched between continents that are now either side of various oceans. There was also clear evidence of glaciers and glacial activity on each of the continents, which showed that they had been grouped together around the South Pole in the late Palaeozoic Era. Apart from Antarctica, all of the other continents that Wegener said formed Gondwanaland – the southern supercontinent – are now positioned around the Equator in tropical, sub-tropical, or temperate climate zones.

If only one or two lines of evidence were known at the time that Wegener proposed the idea of Continental Drift, it might have been reasonable to question whether the different continents could have been joined together as a single supercontinent, but with so many different lines of evidence available, the proposition seems correct. Remember, by then geologists were beginning to think that climatic changes had occurred, and Wegener believed that Continental Drift provided the process by which changes in the relative positions of the continents could account for the distribution of palaeontological and climatic data exhibited by the rocks and fossils. How else could evidence of a glaciation found in India, Australia, Africa, and South America fit together?

He also thought that Continental Drift could account for the formation of mountains. The Americans thought that mountains could only form along the edges of continents through Isostasy, as they were primarily the product of vertical movements. Wegener however recognized that younger mountains tended to be located on the edges of continents, whilst the older ones were often to be found in the centres of continents.

The Americans in particular rejected his ideas in the 1920s and 1930s, because they thought that, even if he had the evidence and an explanation for the distribution of palaeontological and stratigraphic data, as well as the spatial distribution of mountains and apparent structural fit between the southern continents either side of the Atlantic, he did not have a mechanism that could produce the large-scale horizontal movements required for the continents to move around the globe.

Discussion point

Is the lack of a "driving mechanism" a problem in the acceptance of a theory? Many theories in geology have been accepted even though the author(s) did not have evidence of a causal mechanism at the time. You only have to look back at most of the topics presented in this book to find examples of this. Why then did the Americans in particular reject Wegener's theory on these grounds?

American geophysicists thought that gravity and seismic evidence indicated that the Earth was solid, just as Kelvin had proposed. Therefore, they argued that if this was true, how could the continents move? They rejected the idea of Continental Drift because – amongst other things – movement would require a fluid or at least a plastic layer beneath the continents. This is odd given that Isostasy, which was one of their best theories, tended to indicate the presence of a viscous plastic layer beneath the continents. It was a theory dominated by vertical movement that required a plastic layer. They thought that a fluid layer could not exist because seismic waves could not pass through a liquid; in fact, some seismic waves can pass through liquids. At the time, one of most prominent supporters of Isostasy, Chamberlin, had talked of "Isostatic Creep", which resulted in the gradual lowering of continents and the rise of ocean floors and sea level, which then led to the deposition of marine sediments on the continents. This in turn was followed by the sudden rise of the land surface. However, this still implied only limited horizontal displacement, which could not account for the large-scale horizontal movements Wegener was suggesting.

By the 1920s, American and European geologists felt that the "subcrust" had to be mobile under pressure, so that isostatic equilibrium could be maintained. This meant that there had to be a gradual movement of the subcrust from under the oceans towards the continents, or visa versa. Many of these geologists also considered that horizontal motion or "lateral creep" could occur, but the problem was still a question of the scale of movement. This allowed them to accept the theory of geosynclines, which produces mountains primarily through vertical displacement.

Wegener's argument was based on the scale and extent of the mobility. He believed that the continents were "light" – as did Airy's version of Isostasy – and that they became thicker (an average thickness of 100 km), while the oceans consisted of a denser, "heavy" material – as in Pratt's isostatic model, which was no more than 5 km thick. Therefore, the continents could float on and "plough" through the Oceanic Crust. He thought that one of the most important elements of his theory was time: that is, given enough time, the continents could force their way through the "plastic" substrate.

As we have seen above, one of his lines of evidence was the positions of mountains. These tended to be on the western or equatorial sides of continents. The American, Frank Bursley Taylor (1860–1938), thought that equatorial mountains were formed because continents tended to migrate away from the poles; Wegener thought that they moved towards the Earth's centre of gravity.

At the time, Reginald Daly (1871–1957) – a Canadian geologist who became a Professor at Harvard – was one of the few North Americans who seemed to be interested in the concept of Continental Drift. He came up with four explanations for continental movement:

1. *Tidal retardation*, i.e. the frictional effect of the surface area of the seas. Fossil and stratigraphic evidence indicated that in the past seas were more extensive than at present, therefore tidal friction would have been greater and continental movement more extensive. This meant that Continental Drift was variable and dependent on relative sea levels.
2. *Gravitational instability*, due to isostatic readjustment – this could have been greater in the past than at present.
3. *Gravitational effects*, due to the overthrusting of sections of the crust.
4. *Gravitational effects*, due to the irregular shape of the Earth.

Whilst Daly was on an expedition to South Africa, he met Gustaaf Adolf Frederik Molengraaf (1860–1942) – a Professor of Geology at Delft University of Technology in the Netherlands – and Alexander du Toit (1878–1948), considered to be the most important geologist from South Africa. In the early 1920s, Molengraaf had begun to consider the idea that the continents moved around the globe like "ice flows" and that the Indonesian Archipelago was the result of two continents converging. He proposed a mechanism for sea-floor spreading, which accounted for the opening up of the Atlantic Ocean and the East African Rift Valley. Du Toit published two important books, one in 1927 that covered the evidence he had found whilst conducting fieldwork in Africa and South America called, *A Geological Comparison of South America with South Africa*. The other, published in 1937, was titled *Our Wandering Continents; An Hypothesis of Continental Drifting*. By the time that Daly returned to America from his expeditions, he had accepted the idea of Continental Drift. Why? Because he had seen much of Wegener's palaeontological and stratigraphic evidence for himself, and he had been with geologists who had embraced the theory.

Although seismologists had "proved" that the Earth had a zone beneath the crust that was rigid, this did not mean that it had to be solid. Daly named this layer the asthenosphere, which means the "zone without strength". He also argued that in the oceans the colder, denser basaltic Oceanic Crust was in an unstable condition on the lighter, hotter layer that existed beneath it, and that it was only its internal strength and continuity together with the adjacent Continental Crust that kept it in position. He thought that if the Oceanic Crust were to fracture, it would sink and drag the continents down with it. He called this "Gravity Sliding", although the process is now known as the "Slab-pull Model". Daly used this model to account for the "missing" part of the Geosynclinal Model, i.e. the formation of fold mountains. He argued that deposition of large quantities of sediments in the oceans led to downwarping of the ocean floor, which in turn led to heating and melting of the sediments. The downwarping then caused the non-solid substrate to flow outwards towards the continents, which led to the elevation of the continental margins. The thinned and weakened, downwarped Oceanic Crust would eventually fracture and slide down into the mantle due to the force of gravity. Finally, as it slid down, it would fold the overlying sediments. This is essentially the process that operates in subduction zones in the Plate Tectonics model.

Daly's ideas helped to answer some of the problems of Continental Drift, i.e. that if oceans were able to give way to the continents as they "ploughed through them", why did the continents deform. With this concept, folded sediments along their edges were incorporated into, but did not form part of, the original continents. This answered the problem of the observed age difference between the rocks found in fold mountains and those of the adjacent continents. Daly was therefore able to link the two theories of Continental Drift and geosynclines together, which meant that he had linked together a theory that the American geologists did not believe, with a theory they thought was undoubtedly true. However, if they accepted the link, it would mean that they would also be accepting the idea of a layered rather than a solid Earth, a concept most of them had rejected because they still thought that their seismic information indicated that the Earth was solid.

Discussion point

How often have we encountered the idea of a solid Earth in this book? Each time, people have presented it with great scientific credibility.

At about the same time, Irishman John Joly (1857–1933) – a Professor of Geology at Trinity College Dublin, credited with the development of using radiotherapy for fighting cancer – was also working on the idea that radioactivity was the "eternal heat source" for tectonic processes and that molten rocks would disrupt the overlying crust. This implied that the driving force behind the Earth's "cyclic" history of mountain building was the build-up and release of radioactively generated heat. He explained that if the layers of rocks were thick enough, they could generate sufficient heat to make them plastic, which would then allow them to deform. This, he thought, could explain why geosynclinal sediments were always so thick. He also started to consider that heat convection occurred not only in early Earth history, but could be a continuing process (see also Arthur Holmes below). He cited the large-scale flood basalts found around the globe as evidence of this. He also felt that thermal expansion of the basalts as they rose led to their eruption through rifts, which would enable them to spread out over the sea floor or land surface.

Joly began to look at what appeared to be contradictory evidence, such as:

1. Geodetic surveys had proved that sunken continents did not exist.
2. Palaeontological evidence indicated that different continents must have been connected.
3. Ninety-four percent of all recorded earthquakes and volcanoes occurred along the edges of continents and the geosynclinal belts, which indicated that deformation was usually confined to those areas.

In 1926, Joly presented his own tectonic theory in *The Surface History of the Earth*. This was based on the thermal cycle outlined above, in which he linked periods of "revolutions" with known orogenic (deformation) events. He thought that the Earth "leaked basalt", which implied that there was basalt under the granitic crust. Although the idea of molten basalt beneath the crust did not fit with geophysical evidence, Joly argued that it did not have to be uniform in thickness or continuous in its distribution. Originally, he fitted his ideas into the existing Geosynclinal Model, but when the idea of Continental Drift was proposed, he saw how his model helped to explain the movement of "continental rafts".

As quoted in Oreskes book:

> During the period of thermal dissipation the ocean floor will be attacked by hot and, doubtless, often superheated currents from beneath … It is possible that magma currents … may result in the location of fractures principally on the western coasts. The fractures will be rapidly filled in by congealing basalt.

Joly submitted his ideas in a paper to the Royal Society but it was turned down because even though geologists had "enthusiastically received it at the Geological Society", other scientists (probably physicists, according to Oreskes) did not accept it.

Another very important person involved in the development of Plate Tectonics was Arthur Holmes. In 1922, he proposed that the regional-scale overthrusting identified in the Alps by Elime Argand (1879–1940), a Swiss geologist, was a result of the intrusion of molten material from below the continents, as Joly had suggested. This, he thought, provided the lateral stresses necessary to fold the rocks. Between 1927 and 1929, Holmes continued to develop his ideas and proposed the concept of convection currents due to the "differential heating by radioactive decay", which he thought operated in the Earth's mantle. He also thought that ocean basins would form where basaltic magmas erupted to the surface in zones of crustal tension, and that the top of the mantle could not be completely liquid because if it were it would not be able to "grip the overlying continents to be able to drag them". This implied that it was the convection currents operating in the mantle that moved the crust (Fig. 8.9). This was not a new idea, as it dated back to at least 1872.

Discussion point

As Palmer says:

> Now physicists really had to take notice and admit that Kelvin was way off the mark. It began to seem that the geologists had been right all the time to argue that the Earth must be much, much older than 20 million years or so.

By the 1930s, many geologists had embraced the idea of continental movement, because there was a wide range of evidence, which seemed to indicate that it could have happened, and at least three different driving mechanisms had been proposed:

1. Daly's gravity sliding
2. Joly's periodic fusion
3. Holmes' mantle convection.

Although Wegener's ideas were quite widely accepted in Britain and by 1925, some Americans were seriously considering his ideas; the overall reception in America was still negative until the 1960s.

Oreskes points out that American geologists "didn't like the European trend to follow other people's ideas and their excessive acceptance of their

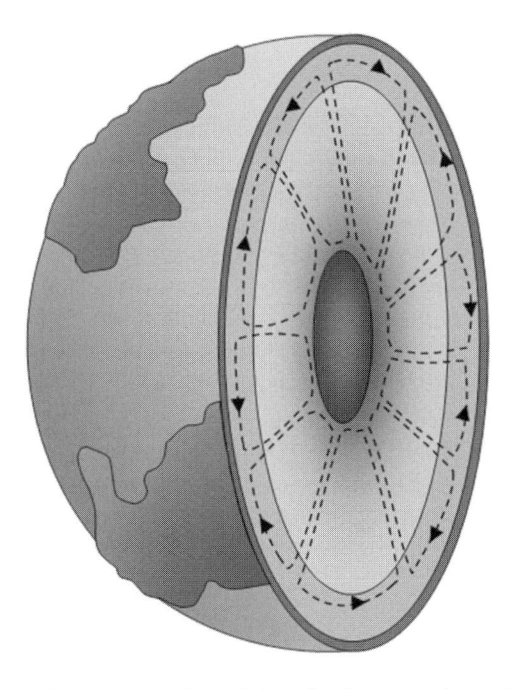

Fig. 8.9 **A schematic representation of the Mantle convection cells**

scientific authority". Therefore they felt that geological processes could rarely be explained by a single "all encompassing idea" or an idea focused on one person's beliefs, which then became self-perpetuating and often (at least in the short term) self-fulfilling. Rudwick's excellent book also highlights the difference between British and continental geological thinking and practices.

Boulter includes an interesting commentary on the approach of scientists from different nations to investigating scientific problems. He uses the analogy of different groups of private detectives all competing to find clues that will solve a case:

> There are international groups, each with their favoured techniques and attitude. The Americans like to check the basic building blocks, the data themselves. The French are very concerned that a logical protocol has been followed and that the analysis of data is properly validated statistically. The English have hunches, a suspicion of simplicity and the desire to test. All three groups are keen to prove the other wrong and maybe won't be too upset if they are eventually proved to be wrong themselves.

This may seem an exaggeration or an extreme, but in general terms the three groups mentioned have followed these three distinctive approaches to science for the last two to three hundred years.

Discussion point

If you think back to many of the ideas outlined in previous chapters, and the way in which subsequent geologists followed either particular ideas or particular people, you can see how geological thought has or has not progressed. Given this information, the American standpoint seems to have been quite reasonable. But were they really treating all theories with "equal weight", or were they tending towards favouring particular ideas or people, just as their European counterparts had done? If you look at the paragraphs below, you can see that in reality they found it difficult to be "open-minded", because they too had strong personalities who held positions of influence with particular views. This automatically leads to the question; can scientists really be truly objective?

What was the basis for being "objective" and why did the Americans feel so strongly about it?

Oreskes explains that the Americans decided that, like many of their social ideas, they should try to develop science independent of their European equivalents and, just like their social philosophy, everyone was equal. They worked on the premise that you collected facts and then developed theories, which was not always the case in Europe. As we have seen in the previous chapters in this book, a number of geologists developed theories without all the evidence required to prove them. Hutton, for instance, effectively developed his theory and then found some of the crucial evidence to prove it. The other basis for their approach to science, according to Oreskes, was their underlying principle that "knowledge is founded on the hard work of compiling carefully observed facts".

Discussion point

The problem we often face is this: where do fact and theory begin and end? How often do you have to have a theoretical model in mind before you can begin to look for evidence? Developing that argument still further, you are rarely able to look for evidence without having some idea on which to base your observations. This is true whether you are working in the field or in a laboratory. It is difficult to start with a "blank page". It is interesting to think back to the work of Alvarez (Chapter 5) when looking for evidence at the K-T boundary.

The Americans therefore generally rejected Wegener's ideas because he did not have all the evidence, particularly a "driving mechanism", before he developed his theory. Unfortunately, this meant that they had to reject a great deal of field-based evidence, a method of evidence-collecting they were particularly keen on, and on which they founded much of their geological knowledge.

The Americans decided that they needed to find their own field evidence to be able to "test" the various theories. However, they initially rejected a proposed field expedition to South America to record similarities between the rocks there and South Africa, because the original proposal had been written in terms of a "test of Continental Drift". The proposal was then rewritten so that the expedition would look more widely at the meaning of the rocks without "a preconceived notion". This work took du Toit two years to complete, after which he concluded that the field evidence suggested that South America and South Africa could not have been more than 400–800 km apart when the rocks were formed. His previous work in South Africa had already led him to this conclusion. He looked at each rock unit independently and came to the conclusion that there had been a Carboniferous glaciation centred in the Atlantic. His comments reveal that he felt that Continental Drift best explained the evidence he had found; for instance, Oreskes quotes the following from him:

> Under the displacement hypothesis, the South African continent is viewed as having proceeded to drift westwards during the latter part of the Mesozoic, crumpling up in its path the marine sediments bordering the Western side of the Brazilian "shield" … The Tertiary foldings and overthrustings, directed usually towards the east, that characterize the Cordillera and extend from Venezuela to Tierra del Fuego, find their explanation in logical fashion under this hypothesis.

However, large-scale continental movement was still not widely accepted in America. For example, William Bowie (1872–1940), an engineer and geodesist, still regarded Isostasy as a local phenomenon, which meant that the Earth's surface was fixed in its relative positions. This implied that the crust was passive and only erosion and sedimentation could change the stresses that led to Isostasy. It also meant that Isostatic movement was a purely vertical phenomenon, as in Pratt's model. Bowie therefore thought that the break-up of South America and South Africa had to be an early geological event. Bowie – who had worked with Hayford, who in turn had worked with Pratt – felt that the Pratt model of a uniform thickness of the Earth's crust was correct and therefore rejected the idea of Continental Drift, as it implied that the crust could vary in thickness.

By 1925, Bowie had changed his mind and thought that the fit between South Africa and South America provided some of the best evidence for Continental Drift, suggesting that a worldwide measurement of longitude by radiotelegraphy would be a good test to find out whether the two continents had moved. The readings were collected together by Gustave-Auguste Ferrie (1868–1932) – the scientist, engineer, and army officer who had developed telegraphy and radio communications in France. The survey, which was conducted in 1933, was compared to a similar survey carried out in 1926; however, the results showed no discernible change in the distances between the two continents. It had failed to show any movement because the limit of error in the measurements was 3 m. We now know that the two continents are only moving apart at a rate of about 25 mm a year, which would only produce a movement of approximately 1.5 m over that time interval.

Charles Schuchert (1858–1942), an American palaeontologist famous for his work in palaeogeography, also rejected Continental Drift because it did not fit with the principles of uniformitarianism and a steady-state Earth. It is interesting to see that he rejected it, even though much of the evidence and interpretation that Wegener had used to develop his theory was palaeogeographic data.

Some Americans had begun to think about "non permanent continents and oceans". For example, Joseph Barrell (1869–1919) – the American geologist who proposed that sedimentary rocks were produced by the action of rivers, wind, and ice as well as marine processes – also questioned the "local and complete nature of Isostasy". He thought that the continents had a weak substrate, which led to basalt intrusions and the fragmentation of land masses.

Other American geologists continued to think about the problem of mountain building and the relative distribution of marine sediments. It appears that one of the reasons that they maintained their views with regard to Isostasy and its incompatibility with Continental Drift, was because most of their original ideas were based on studies of the Appalachians. When they began to investigate the Rockies and other fold mountains, they found evidence of large-scale folding, nappes, thrusting, and crustal thickening – all of which tended to indicate, as European studies had shown, that the crust had a non-uniform thickness and that the formation of fold mountains involved large-scale horizontal movements (Fig. 8.10). The Rockies also provided evidence that there was land where there had once been sea. Holmes wrote to Schuchert, saying that he was certain that it "was impossible to get rid of lands that formerly occupied the sites of present oceans except by moving them sideways".

Fig. 8.10 **Two photographs of intense folding exposed in the cliffs at Crackington Haven, Cornwall (left), and Back Bay, Southern Uplands, Scotland (right)**

Discussion point

As Oreskes says, "Schuchert – who was one of the most influential geologists in America at the time – was running out of alternative ideas to Wegener". He turned back to the old idea of land bridges, because of his conviction that the continents were permanent features, which was a consequence of his view of Isostasy.

In the 1930s, Beno Gutenburg (1889–1960), the German-born seismologist who worked with Charles Francis Richter (1900–1985) to develop the Richter Scale, thought – through his studies of deep-seated earthquakes – that large-scale continental displacements were possible. Guttenburg was famous for the study of the Core-Mantle boundary, which is generally known as the "Gutenburg Discontinuity".

Another line of evidence, which would eventually be used to confirm Continental Drift, was the measurement of gravity at sea. Isostasy was a theory based on gravity measurements on the land, but similar measurements across the oceans had been impossible until Felix Andries Vening Meinesz (1887–1966), who became Professor of Geodesy at the Delft University of Technology, built a gravimeter that could be used at sea. It was installed in the Dutch submarine K-XIII, and was used to undertake a worldwide survey that included the Java Trench. The American Frederick Eugene Wright (1877–1953), a famous petrologist from the Carnegie Institute, also devised a similar instrument; this meant that for the first time the gravitational effects of Isostasy could be tested beyond the edges of the continents. If Isostasy was in balance – i.e. in equilibrium as the Americans thought – it would mean that differences in gravitation should only be found in areas where very recent changes could occur due to large-scale sedimentation (such as deltas). It would also imply that there should be no gravitational anomalies in the oceans, which were considered to be ancient features that were in isostatic equilibrium under the Permanency Theory. But Meinesz's work suggested the opposite – that there were no gravitational anomalies over the Nile Delta, but a large negative one existed under the Java Trench.

A second survey in the submarine S-21 was planned for the West Indies and the Mississippi Delta. Again, the results showed that the predicted positive gravity anomaly did not exist over the delta, but there was a large negative anomaly over the Bartlett Deep. They also found that this anomaly projected beyond the ends of the trench into what appeared to be a "normal ocean floor". The Gulf of Mexico produced another problem: with its known high sedimentation rates, the gulf had a large positive anomaly but there was no discernable topographic feature associated with it.

The basic problem was that some of the data implied that areas of the Earth were not in isostatic equilibrium. They then began to consider the idea that isostatic compensation was restricted to a plastic substrate and that the ocean floor had a different density to that of the continents. This conformed to the Airy Model and the idea that continents might not have a uniform thickness and that they existed above a denser substrate. Gravitational data also indicated that regional stresses existed in the Oceanic Crust, which meant that they could not be part of the "weak substrate" either. This implied that the continents could not move around on the Oceanic Crust and that equally, they could not "plough through" it, as Wegener had proposed.

Further submarine expeditions were conducted in 1932 and 1937 to the East Indies, Haiti, and Cuba. One of the people on the first trip was Harry Hammond Hess (1906–1969), who later joined Princeton University, where he developed his theory of Oceanic Ridges and sea-floor spreading. Whilst sailing back and forth across the Pacific Ocean, Hess collected a significant body of sonar data, from which he developed his ideas. William Maurice Ewing (1906–1974) went on the second trip. Ewing was an American geophysicist and oceanographer who developed the use of seismic reflection and refraction for use in studying ocean floors and subsea structures. He later became Professor of Geology at the University of Columbia and the first Director of the Lamont–Doherty Earth Observatory.

Both of these trips confirmed the previous results that the Caribbean Plate (as it has been named) was moving eastwards and that the Bartlett Deep was a pull-apart structure on the edge of the moving block. Hess also noted the association between the Java Trench and the locations of earthquakes and volcanoes that occurred at a depth of approximately 140 km on the landward side of the trench. He thought that the position of the negative gravity anomaly associated with the trench, the earthquakes, and the volcanoes all indicated that this was an area of compression and melting.

In 1933, Hess introduced the concept of a "Tectogene", which is an area of downwarping crust. He also thought that there could be areas of asymmetric folding such as nappes, which were formed due to the overturning and fracturing of downwarping rocks. This, he proposed, resulted in distinctive belts of horizontal compression on a global scale.

Laboratory studies that looked into the processes of crustal deformation were also being conducted by Philip Henry Kuenen (1902–1976), a Dutch geologist at the University of Groningen, and David Tressel Griggs (1911–1974), who studied the mechanical properties of rocks at high temperatures and pressures at the University of Los Angeles. Both were trying to model underthrusting of the crust using two different processes: Kuenen used compression and Griggs used rotating drums to induce viscous drag. During Griggs' experiments, he observed that drag also led to the piling up of material around the area where down-folding occurred. He used this to suggest that there could be a convection cell under the Pacific Ocean, which resulted in the sinking of material near the mountains around the edge of the Pacific. He also suggested that this would produce the deep-seated earthquakes on plains tilted at 45° towards the continents, which Gutenburg and Richter had previously discovered. This plain was later named the Wadati–Benioff Zone after the two scientists who first proposed its existence. It is interesting to note that even though Wadati's paper was published before Benioff's, it was written in Japanese and consequently his name is usually missing from the name – hence it is more generally known as the Benioff Zone.

In 1937, Nicholas Hunter Heck (1882–1953), a civil engineer and seismologist, presented a map at the *Symposium on the Geophysical Exploration of the Ocean Bottom*, which correlated earthquakes with the Mid-Atlantic Ridge, a structure that he thought might be linked to similar features found in other oceans. It has been suggested that this symposium signalled the end of the line for a number of "well established theories". It included reviews of the thickness and uniformity of the crust, its composition, and the idea of remnant magnetism. It showed that there was clear evidence:

1. That ocean basins were not static;
2. Of the existence of large-scale horizontal compressive stresses in the crust that were involved in crustal down warping, earthquakes, and volcanoes;
3. That there was also a theoretical model for convection currents in the mantle and crustal drag, which could account for all of these.

Discussion point

It is interesting to note that Hess, Menesz, Kuenen, and Griggs all thought that crustal thickening was due to the presence of crustal roots – which was part of the Airy model – rather than underthrusting, which is the present view.

In 1937, John Adam Fleming (1877–1956), a civil engineer who became a geophysicist at the Department of Terrestrial Magnetism of the Carnegie Institution in Washington, noted that some rocks showed a reversed magnetic polarization and that others showed different relative positions for the magnetic poles (Chapter 2).

Scientific developments during the Second World War also led directly to large-scale gravity, magnetic, temperature, depth, and acoustic surveys – all of which were regarded as classified information by the Americans. At the same time, British geophysicists were working with similar unrestricted data.

Stanley Keith Runcorn (1922–1995) – an English geophysicist at the University of Cambridge – felt that if this magnetic data were unreliable, it would indicate a random distribution of magnetic data, but that if it were accurate, there should be a pattern to it. He produced a set of "apparent polar wandering curves" for the UK, the first of which was published in 1954. These showed changes in the position of the UK from the Precambrian all the way through to the Tertiary. He thought that they showed that the magnetic poles had moved; he was also the first person to discover that the Earth's magnetic poles had periodically reversed (Chapter 2).

What did the polar wandering curves represent? Had the poles moved or had the continents moved? Runcorn and Edward Irving (born in 1927) – who was also at Cambridge – proposed a solution: if data were collected for a number of continents, the results could be compared. They thought that if the data showed a consistent pattern, it was likely that the poles had moved, but if the data were random, it was more likely that the continents had moved.

Irving collected data from Australia, India, North America, and Europe. The results showed paths that were different but consistent with the drift positions suggested by Wegener's palaeoclimatic data. By 1956, Irving and Patrick Maynard Stuart Blackett (1897–1974) – who later became a winner of the Nobel Prize for Physics for his work on cosmic rays – were arguing that palaeomagnetic data was evidence for Continental Drift.

This period marked the start of a rapid increase in investigations of the oceans, the ocean floors, and the Oceanic Crust (Fig. 8.11).

Fig. 8.11 **Two views (top left and right) of the foreshore at Coverack and Porthoustock (lower left), both in Cornwall, and Grey Hill, part of the Ballantrae Ophiolite Complex, at Girvan, Scotland (with Ailsa Craig in the background (lower right))**

In 1957, Walter Heinrich Munk, at the Scripps Institute of Oceanography in California, proposed the idea of drilling down to the "Mohorovicic Discontinuity" – the boundary between the Earth's Crust and the mantle; this became known as Project Mohole. (The Mohorovicic Discontinuity, which is usually known as the Moho, is named after Andrija Mohorovicic, the Croatian geologist who first proposed its existence in 1909.) This was supposed to be undertaken in three phases, the first of which was to drill five holes – the deepest of which cut through 601 feet (183 m) of oceanic sediments and crust in 11,700 feet (3,500 m) of water off the coast at Guadalupe, Mexico. In fact, it penetrated 557 feet (169 m) of Miocene sediments that lay on top of 44 feet (13 m) of basalt. Having achieved this, the other two phases of the project were subsequently abandoned in 1966.

Discussion point

In his book, *Principles of Terrane Analysis*, David Howell explores the idea that significant age and density differences existing between the Continental and Oceanic crusts were partially responsible for the failure of the Moho drilling project. He says that:

> This notion contributed to the rationale for the ill-fated Moho drilling project because they believed that a single deep borehole through the strata above the Oceanic Crust would sample horizons going back to the very beginning of the rock record [because] as recently as 1950s people thought that the more mafic and dense simatic crust of ocean basins represented the oldest crust.

Given the relative inaccessibility of the Oceanic Crust, it is still fascinating to think that you can walk across the Moho as you walk along the beach at Coverack, and across part of a sheeted dyke complex at Porthoustock, both in Cornwall, or an Ophiolite Complex at Ballantrae, Southern Scotland (Fig. 8.11).

8.5 Plate Tectonics

In 1960, Harry Hess proposed that convection currents were the driving mechanism for Continental Drift; he published his ideas in 1962 through the Geological Society of America under the title, *The History of Ocean Basins*. His proposal concentrated on evidence from the mapping of oceanic ridges, which had high heat flows and down-faulted valleys along their centres. There also appeared to be very few features in any of the oceans that were older than 100 million which, when compared to the average age of the Continental Crust, was remarkably young.

Discussion point

Hess is usually acknowledged as the person who worked out the process by which Oceanic Crust is generated but, as Oreskes points out, he did not acknowledge the work of Holmes or any of the other people who had already proposed that convection currents in the mantle were the driving mechanism for Continental Drift and Plate Tectonics (see above).

She therefore suggests that Hess' work was not revolutionary, as it is usually portrayed but was evolutionary – in other words, it was a development or a confirmation of previous ideas. After all, Holmes had said in 1945 that Oceanic Crust would be consumed by sinking back into the mantle where convection currents descended and that new Oceanic Crust would be generated at Mid-Ocean Ridges. She also says that Hess actually rejected the idea that a physical process could generate Oceanic Crust, as he thought that it had been formed by a chemical (hydration) reaction through contact with water.

In 1961, Robert Sinclair Dietz (1914–1995) – Professor of Geology at the Arizona State University, who had been working with Hess – proposed a mechanism for the production of Oceanic Crust and introduced the term "Seafloor Spreading". He had been working at the Scripps Institute of Oceanography and had been studying the Emperor Chain of seamounts that extend north-westwards from the Hawaiian Islands. As early as 1953, he had considered the idea that these volcanic islands may have formed as the seabed, on which they existed, moved north-westwards, as though they were on a conveyor belt.

Whilst working on the Tertiary basalts of the Snake River Plane, in America, Alan Cox (1926–1987) found magnetic reversals within particular layers that appeared to be internally consistent. In 1963 he, together with Richard Doell and Brent Dalrymple, both at the time working with the US Geological Survey, published the first geomagnetic time scale. This was then compared with the offshore magnetic data and both appeared to tell the same story. They showed that the Earth's magnetic poles had frequently changed their polarity in the past, but it also indicated that these changes were complicated and irregular, the causes of which are still being debated today.

The hypothesis of magnetic stripes in the sea floor was also independently proposed by Lawrence Morley, a Canadian, and Frederick Vine and Drummond Matthews, both English marine geologists and geophysicists at the University of Cambridge. It is interesting to note that Morley's paper was rejected for publication, whereas Vine and Matthews' *Magnetic Anomalies over Ocean Ridges* was accepted by the magazine *Nature*. By 1966, Vine, Matthews, and two Americans geophysicists at the Lamont–Doherty Geological Observatory had all proved the existence of magnetic stripes. Neil Ordyke (also at Lamont–Doherty) also proved the link between the magnetic stripes in offshore sediments and onshore basalts, which demonstrated that both onshore and offshore rocks showed the same data and therefore produced a unified record. Dan MacKenzie at Cambridge and Jason Morgan at the University of Princeton independently looked at the relationships between the magnetic stripes to try to determine the relationships of the movements. Finally, Morgan proposed the basic idea of Plate Tectonics in 1968, and Vine and Matthews were the first people to show a correlation between oceanic magnetic stripes and the Geomagnetic time scale produced by Cox in 1969 (Fig. 8.12).

John Tuzo Wilson (1908–1993), a famous Canadian geologist and geophysicist at the University of Toronto, showed an age relationship between the Hawaiian Islands and suggested that their presence was due to movement of the Oceanic Plate on which they were formed over a hot spot. In

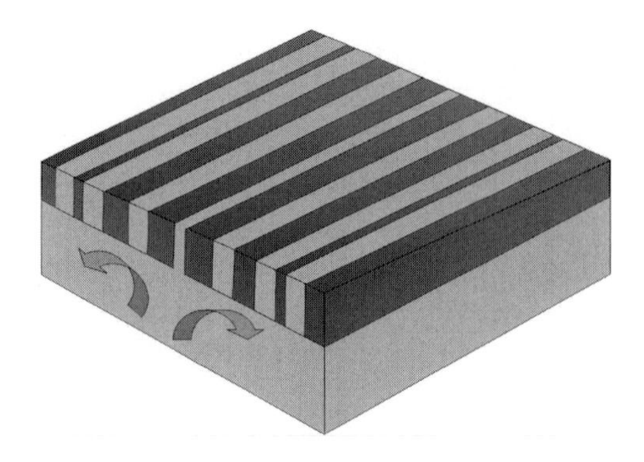

Fig. 8.12 **A diagrammatic representation of the formation of the magnetic stripes in the ocean floor, where the darker bands represent positive (N-S) orientations, and the lighter one, negative (S-N) bands**

1965, he also proposed the idea of transform faults, collision boundaries, and spreading ridges.

Oreskes, together with many other authors, present the evidence for Plate Tectonics, which was different from that for Continental Drift. This included:

1. Land-based palaeomagnetic measurements, which showed the divergence of polar wandering paths for different continents with increasing geological age;
2. Marine palaeomagnetic stripes, which confirmed the hypothesis of sea-floor spreading;
3. Detailed seismological evidence, which showed that Oceanic Crust splits at ocean ridges and sinks at ocean trenches;
4. Plate movements and rotations around fixed poles.

This differed from Wegener's evidence for Continental Drift, which included:

1. Palaeontological evidence, which indicated the movement of species between continents during specific geological periods;
2. Consistent stratigraphic sequences across a number of continents;
3. Palaeoclimatic indicators, which could not be explained by simply changes in local climates;
4. The jigsaw puzzle fit of the continents.

Although most of Wegener's data had been recorded in the field through direct observation of rocks and fossils, the evidence for Plate

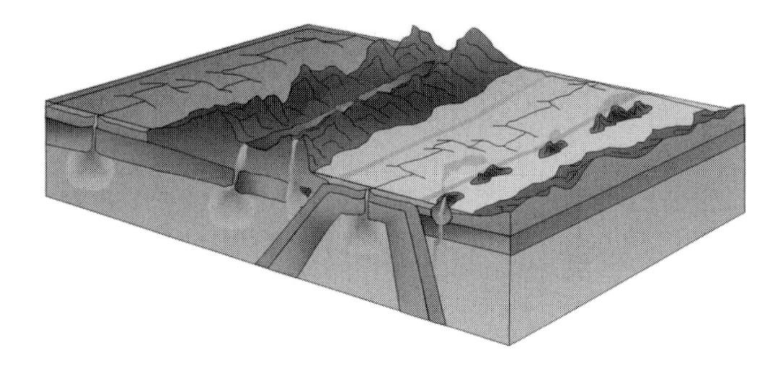

Fig. 8.13 **A diagrammatic representation of the Plate Tectonics model**

Tectonics was based on geophysical and other indirect data (Fig. 8.13). As Oreskes says:

> The advocates of plate tectonics seldom invoked the older evidence in support of the new theory, although many were aware of it. The stratigraphic similarities, the fossil resemblance's, the palaeoclimatic data, played no important role in the establishment of plate tectonics. Rather, the old homologies were used *ex post facto* as a demonstration of the far-reaching explanatory power of the new theory.

It is clear that the theory of Plate Tectonics actually validated Wegener's evidence, and that since the 1970s, geologists have used Wegener's methodology, based on traditional techniques and lines of evidence, to explore the inner workings of much of the Earth's surface.

Discussion points

From the above, you can see that the evidence that was used to propose and prove the idea of Continental Drift was geological, whereas the data for Plate Tectonics was primarily geophysical. Virtually no geological data was used in establishing the theory of Plate Tectonics, but this theory then proved the reality of Wegener's Continental Drift. Geophysicists proved that many of Wegener's ideas were correct at the time that he proposed them – even though he did not have a driving mechanism for them.

Why do you think that geophysical data was accepted in America rather than geological data that was based on the methodologies American geologists were familiar with?

Could this be another example of the views of Gould, concerning the relationship between geology and the other "higher sciences"?

Oreskes suggests that this problem goes back to the 18th and 19th centuries, when geologists and physical scientists had different approaches to their science: the former worked primarily in the field and the latter worked in the laboratory (see Chapters 1 and 2 for the differences between the ways in which they tried to establish the age of the Earth). Field geology looked (and still looks) at the physical evidence in the rocks and fossils and is therefore observational, qualitative, and inductive, whereas physics, mathematics, chemistry, and geophysics are primarily quantitative. Some people, such as Holmes, managed to work across the boundary between the two approaches – in his case, it was between geology and physics.

Isostasy had strong field-based geophysical measurements in which geologists could incorporate the methodologies of the physicists into their own work, in order to strengthen its credibility. It was also collected using instruments that were therefore independent of interpretation and other's influences and used well-known, well-proven physical principles. This indicated a significant shift in approach and methodology away from traditional, field-based, deductive geology, which was not regarded as "proof" to geophysicists who used "hard science". There was also the issue of looking at all theories equally. In reality, however, the Americans leaned towards one particular theory – Pratt's version of Isostasy – that meant that they had to discount other lines of evidence.

Even though Plate Tectonics was becoming widely accepted from the 1960s, *Geologist and Ideas: A History of North American Geology*, published by the Geological Society of America to celebrate its centenary in 1985, included an interesting contribution by Dwight Mayo. He comments that while some people would argue that geosynclinal theory should have been consigned to the "theoretical 'junk heap'", others still held the view that it was still "the single most unifying concept in geology".

Thus, in spite of the increasing volume of evidence indicating that Plate Tectonics theory represented a significant improvement over the concept of geosynclines for the formation of fold mountains, some geologists still held on to old ideas.

The idea of crustal recycling through spreading ridges and subduction is interesting. As David Howell observes, precise mapping of magnetic lineations indicates that the ocean floor has been recycled, on average, every 110 million years since the beginning of the Palaeozoic Era (Fig. 8.14). As some areas of the Continental Crust are at least 3,900 million years old, this means that the Earth's oceans could have been formed and destroyed 34 times over the same period. Howell estimates that this would have involved recycling 7% of the Earth's mantle, producing an area of Oceanic Crust equal to $10.5 \times 10^9\,\mathrm{km}^{-2}$, which would be nearly 21 times the surface area of the Earth.

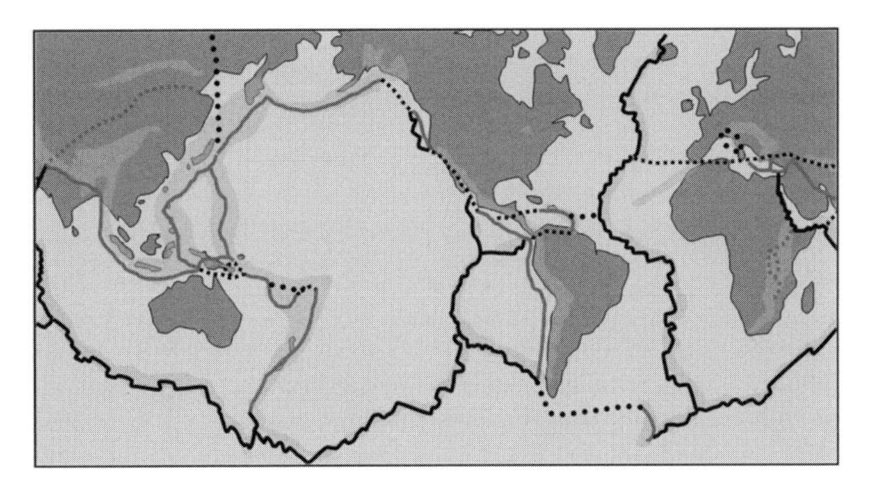

Fig. 8.14 **A simplified map of the tectonic plates. The solid black lines represent the locations of spreading ridges, the grey lines are subduction zones, the black square dotted lines are transform boundaries, the grey square dotted lines are collision boundaries, and the large black dots represent uncertain boundaries**

Further reading

Before we leave Plate Tectonics, it is worth reading the explanation provided by Paul Garner in his book *The New Creationism: Building Scientific Theories on a Biblical Foundation*. In chapter 13, entitled "Global Catastrophe", he outlines the theory of Catastrophic Plate Tectonics. This, like much of the book, provides a clear and balanced overview of many geological phenomena. He then presents an extraordinary alternative explanation/re-interpretation of Plate Tectonics, in which much of the accepted evidence is condensed down into six days of creation, the Genesis Flood, and a 6,000-year Earth history, without explaining all of the implications and consequences that such a system would generate.

Discussion point

It is worth thinking about the following questions:

What were the weaknesses in the American, European, and British approaches to science?
Is it easy to study any area of science in a purely objective way, or is it necessary to have preconceived ideas, theories, or plans?
This highlights how we construct our pool of knowledge – so, can we really be open-minded?
How permanent are the theories we have at the moment?
How may they change in the future?

9
What have we learnt?

One of the most obvious questions that anyone could ask about the history of geology would be – why was Britain so deeply involved in formulating so many of the foundations of modern geology? Geology students are taught that the UK has the most diverse geology for its size of landmass in the world, but would this be sufficient reason to inspire so much activity?

In order to try to understand why Britain may have had such a significant role in the development of so many of the basic precepts of modern geology, we need to look at the wider historical context of the late 18th and early 19th centuries.

The Industrial Revolution began some time between the 1760s and 1780s, and reached its peak around the 1830s to 1840s (the timing depends on which sources you read). Prior to this, ore and minerals had been mined or quarried for a great deal of time, but the introduction of industrial-scale production necessitated a step-change in the way in which these resources were extracted from the ground and transported to the new factories. The growth of Britain's industrial cities, and the prospect of higher wages and better living conditions, attracted huge numbers of workers from the countryside, necessitating the building of new factories and houses, and other amenities. All of this activity relied on locating and extracting geological resources such as coal, iron, and building materials.

The growing importance of geology during this period of major economic and industrial change probably helped it to develop into a separate science. We have seen that some of the key geologists, such as Murchison, were concerned that people were wasting huge sums of money trying to find coal and other industrial minerals in inappropriate rock sequences.

Although there were extensive coal seams close to the surface west of the limestone escarpment in the northeast of England, they had largely been worked out by the early part of the 19th century, and it was generally thought that the coal seams stopped at the edge to the limestone

Time Matters: Geology's Legacy to Scientific Thought, 1st edition. By Michael Leddra.
Published 2010 by Blackwell Publishing Ltd.

escarpment. However, because of his understanding of geological sequences, William Smith convinced local mine owners in County Durham to dig the first deep mine shaft through the Permian Magnesian Limestone in search of coal. As related in my book, *Turn and Burn: The Development of Coal Mining and the Railways in the North East of England*, this opened up the concealed Durham coalfield. It also led to the development of the first purpose-built, fully steam-powered railway in the world, from Hetton-le-Hole to the coal staithes at Sunderland. This was then used as the blueprint for the Stockton and Darlington Railway – and, as they say, the rest is history. The rapid growth of the railways also enabled geologists to travel longer distances in relative comfort.

It is obvious that there were a huge number of people working in practical geology – particularly in mining and quarrying – which added a great deal of geological knowledge and understanding that the professional geologists tapped into. This becomes clear when you look at the number of local mining terms and names that are still buried in everyday geology. We saw in Chapter 2 that in Europe, Johann Gottlob Lehmann, Giovanni Arduino, and Abraham Gottlob Werner each developed their rock classifications to assist in predicting where industrial minerals and rocks could be found.

The effect of the Industrial Revolution resulted in enormous changes that spread out rapidly from Britain to Europe and other parts of the world. It was during this period that both Hutton and Lyell produced their most important publications. The Industrial Revolution also coincided with the period in which the majority of the geological time periods and systems were established and the Geological Society of London was formed as the first such society in the world. At the same time, William Smith drew his famous geology map of the UK; Richard Owen invented the term "dinosaur"; and Charles Darwin set sail on the *Beagle*, leading to the development of his theory of evolution and a significant change in the study of biology.

But what else was going on? Britain was at war. The country was involved in the American War of Independence, a war with Holland and yet another one, the Napoleonic War with France. Queen Victoria came to the throne and the French Revolution caused upheavals in that country that sent out ripples of consternation throughout Europe and Britain. This period marked the transition from the First British Empire to the Second, as Britain turned its focus away from America towards Asia, Africa, and the Pacific. It also signified the start of the Imperial Century in which the UK held sway over a significant proportion of the globe, giving us access to new rocks and resources. Another important social, political, and economic milestone was the passing of the Anti Slavery Act that outlawed slavery in Britain and her colonies. Even this had an impact on geology, as Henry de La Beche for example, had to start working for the Geological Survey, as financial support was reduced from his family's sugar plantations.

Throughout this book, it is clear that most of the 18th- and 19th-century geologists were still largely natural philosophers working in a wide range of subject areas, and were often either serving members of the Church or had close links to it. For many of them this had a major influence on the way they approached geology. There is also some indication that the British view of their European counterparts was coloured by differences in both political and religious backgrounds, which – combined with the upheavals that were occurring across the continent and various wars – meant that they were often isolated from continental ideas or treated them with caution. We saw the effects of this with Richard Owen's promotion as an English equivalent to Georges Cuvier or Roderick Murchison's "land-grab" for Britain. We also saw that continental geologists were more ready to accept the notion of extinctions and other catastrophic events than their British counterparts, who had fallen under the influence of Lyell and the uniformitarianists. As we have also seen, there were, and probably still are, fundamental differences in the way in which science operates in different countries, which affects how scientists view science itself.

Since the 19th century, numerous geologists have built on the foundations established during that crucial period, allowing geology to remain an interesting and dynamic subject. Although it has had a long history, it has gone through significant changes in the last 50 years, in particular with regard to the technological revolution. We now have equipment that allows us to investigate things that not so long ago were completely unimaginable. Technological advances are occurring so rapidly that it is often difficult to keep pace with them. The move to embrace technology may also be, in part, a conscious or subconscious step towards the "higher sciences" to gain credibility. This may make me sound like a bit of a Luddite, but nothing could be further from the truth, as I spent many years experimenting in high-pressure sedimentary rock mechanics. We should use all the technological advances we can, but we should not lose sight of what makes geology different from the other sciences. I remember hearing that when a physicist took over as the head of a university geology department, he decided to throw out all books that were over 10 years old, because they would be out-of-date. Members of his department were horrified and hid all the classic palaeontological treaties away for fear of losing them. They then reappeared some time later when his views had changed.

The move to embrace technology has also helped to develop different branches of geology, resulting in the geologists who work in them becoming increasingly specialized. In general terms, we have lost the people who have an overview of our science, and have often become so focused on a small part of geology that we end up not being able to see "the wood for the trees". Here is a simple example: I sat in on a series of seminars in which a group of world experts put forward their ideas on a particular subject. I remember thinking that if they had said the same thing to scientists from a different subject, they would have been laughed at. I also heard about a field trip in the Pennines in which someone spent the day pointing out "glacial features" that were clearly the result of lead mining and stone quarrying.

We have seen that geology has had its fair share of internal politics, egos, and red herrings, which have all helped to shape the subject as we know it today. We have seen examples of people and ideas that have been largely ignored by the establishment, because they did not fit in. We have seen examples in which an individual's influence and reputation was so great that it outlived its usefulness, often in the face of overwhelming evidence. We have also seen examples where schools of thought have been self-perpetuating well beyond their sell-by dates. It would be naive to think that science has moved beyond this type of behaviour and that therefore this does not happen any more.

What of the future? Will geology continue to live in the shadow of the other sciences?

In the prologue to his book, *A Crack in the Edge of the World*, Simon Winchester includes the following comments:

> ... like alchemy and the medicine of the leech and the bleeding rod, the Old Geology is a science born long ago (most formally in the 18th century), one that, unlike so many of its sister sciences – chemistry, physics, medicine and astronomy – never truly left the age of its making.

Winchester tells us however, that with recent scientific developments, "never before has any long-existing science been remodelled and reworked so profoundly, so suddenly and in so short a time".

He goes on to say that, following the intellectual revolution of the 1960s, geologists realized that up until that time, they had "quite literally, only been scratching the surface" and that "we had never considered the earth as it truly deserved to be considered". Nowadays the world is viewed as a single system, a concept which is at the heart of the "New Geology".

So, have we produced all the big theories, or are there still some breakthroughs yet to be made?

In this book we have seen a number of the world's best scientists come up with theories that they were convinced would last, but in each case as science gathered more information their ideas and theories were adapted, superseded, or abandoned. How sure can we be that this will not happen in the future? Many researchers might still say that we now know many of the answers, but how often has that been said in the past, until new evidence comes along to change people's views?

It is clear from the different topics covered in this and similar books, that people have built and continue to build great reputations based on their work. Sometimes they then go to great lengths to defend their ideas, even when they are proved wrong. Nowadays, geology, like all the sciences, generally involves big money, which means that there is often more than just reputations at stake in defending your corner. We will continue to see people rise and fall, and theories come and go. It is therefore important to learn from the past, as the past is not only the key to the present but also to the future.

A few years ago, there was a big story in the geological literature and news, which said that a volcanic eruption in the Canary Islands would dislodge a 12-mile (19.3-km) wide fault-bounded block. This would trigger a tsunami that would travel across the Atlantic in nine to twelve hours. It was predicted that when the wave, estimated to be 165 ft (50 m) high, hit the east cost of America, it would travel for a distance of up to 200 miles (322 km) inland. This doomsday scenario received a great deal of publicity (including a TV documentary) and there was even talk of a film based on the idea. Geological folklore has it that when it was presented to a student group, one of the students pointed out that surely the tsunami would lose most of its energy crossing the Mid-Atlantic Ridge. It was quickly realized that the entire catastrophic element of the theory fell apart. Was this a case of the "king's new clothes", or did the publicity machine take over? Alternatively, was it just that most geologists were not interested enough to think through and question the idea; or were they just swept up in a wave of enthusiasm? This example shows that we can still get things wrong, take an idea too far, or at least not question an idea enough before going public. It is comforting to know that, generally however, the peer review system, and the scientific process filters out errors.

It is important when we read or are told anything, that we do not accept things at face value. We must always distinguish between fact and interpretation: there is a huge difference between the two that can change over time. We need to be certain about what we know and where that knowledge comes from, in order to be able to build on it in the future. As the final section in Chapter 5 indicated, the role of a scientist, or anyone remotely interested in science, is to ask questions and not necessarily accept the first answer they are given.

Fig. 9.1 **The Howick Bay Fault, Northumberland**

Alvarez's comments in Chapter 5 and Manning's in Chapter 7 indicate that it is often easy to find what you are looking for, even if it does not necessarily exist.

It is also important not to fall into this trap. I can recall a memorable field trip where a colleague from another subject had been told that a set of important footprints existed at Howick Bay, Northumberland and, sure enough, they found them even though they were looking in the wrong place. The person compounded the error by announcing that they were dinosaur footprints, even though the rocks were Carboniferous in age (Fig. 9.1). Nevertheless, in their determination to find the footprints, they missed one of the most important features and the primary purpose of our visit, the Howick Bay Fault, and failed to recognize it even when standing next to it. The lessons are clear, do not miss the woods when looking for the trees, and sometimes a little bit of knowledge can be a dangerous thing.

At the moment we hear, see, and read a great deal about climate change. Huge sums of money are involved in every side of the argument and, just as the examples above show, there are big reputations being made and defended. Large-scale industrial and national interests all want a slice of the climate change cake and this can affect what is being said. With so much money at stake, it is not surprising that each group will try to steer the science, debate, and conclusions in the direction that best suits their own goals or that has the least effect on their own finances, plans, or research. This may seem a bit of a cynical view but, together with the discussion of catastrophes and the nature of science in Chapter 5, as well as other comments included in Chapter 8, it all comes back to the essential question: can science ever really be impartial or independent?

We may feel that we have so much information available to us today, that the development of new theories are less likely. However, it is clear that others are attempting to re-interpret geology for their own ends. We can ignore them as irrelevant, illogical, or just misguided, but if the public read and accept such ideas without comment or correction from the professionals, we are not only doing a disservice to our subject but to society as a whole, with potentially dire consequences.

On a lighter note, one of the good things about our subject is that there are still things to discover at a range of scales, from the big theories to new fossils, and fortunately there is room for everyone. Douglas Palmer includes the phrase "only scratched the surface" used by Winchester (see above) before explaining that:

> There is plenty of scope for new generations of enthusiasts ready to make their mark on our understanding of the rock record. New finds, new techniques of investigation and even just taking a new critical look at received wisdom can produce startling results.
>
> Fortunately, unlike so many of the pioneers, you no longer have to be a wealthy gentleman amateur or clergyman to pursue the investigation of Earth Time – the future challenge is open to all. Good hunting.

Finally, a message to all geology students, past, present, and future, and those outside academia who have a love for geology – the future of your subject is in your hands. You should treat it with the love and respect it deserves, do not take anything for granted, and be prepared to argue your case based on firm foundations. Also, do not just read the textbooks. You will find a great deal of other information, which will help you understand your subject, on the popular science shelves of bookshops, and they may give you an insight into aspects of the subject that textbooks do not necessarily cover. But a word of warning: you may find that these types of books are a bit like buses – you can wait for ages before one appears, and then a number come along at the same time.

Bibliography

The following is a list of some of the texts used in the preparation of this book. Those marked with an * I would recommend for further reading:

*Ager, D.V. (1991) *The Nature of the Stratigraphic Record*, 2nd edn, John Wiley & Sons, Ltd, Chichester, p. 122.

*Ager, D.V. (1993) *The New Catastrophism: The Importance of the Rare Event in Geological History*, Cambridge University Press, Cambridge, p. 231.

Albritton Jr, C.C. (1989) *Catastrophic Episodes in Earth History*, Chapman & Hall, London, p. 221.

Alexander, D.R. (2008) *Creation or Evolution: Do We Have to Choose?* Monarch Books, Oxford, p. 382.

Allen, K.C. and Briggs, D.E.G. (1989) *Evolution and the Fossil Record*, Belhaven Press, London, p. 265.

*Alvarez, W. (1997) *T. rex and the Crater of Doom*, Princeton University Press, Princeton NJ, p. 185.

Aughton, P. (2003) *Newton's Apple: Isaac Newton and the English Scientific Renaissance*, Weidenfield & Nicolson, London, p. 215.

*Baxter, S. (2003) *Revolutions in the Earth: James Hutton and the True Age of the World*, Weidenfeld & Nicolson, London, p. 245.

Begon, M., Harper, J.L., and Townsend, C.R. (1990) *Ecology: Individuals, Populations and Communities*, 2nd edn, Blackwell Scientific Publications, Boston, p. 945.

*Benton, M.J. (2003) *When Life Nearly Died: The Greatest Mass Extinction of all Time*, Thames & Hudson, Ltd., London, p. 336.

*Boulter, M. (2002) *Extinction: Evolution and the End of Man*, Fourth Estate, London, p. 210.

Bowden, M. (1982) *The Rise of the Evolution Fraud. Sovereign Publications*, Bromley, Kent, p. 227.

Bowler, P.J. (1992) *The Fontana History of the Environmental Sciences*, Fontana Press, London, p. 634.

*Bowler, P.J. (2007) *Monkey Trials & Gorilla Sermons: Evolution and Christianity from Darwin to Intelligent Design*, Harvard University Press, Cambridge, MA, p. 256.

Time Matters: Geology's Legacy to Scientific Thought, 1st edition. By Michael Leddra.
Published 2010 by Blackwell Publishing Ltd.

*Cadbury, D. (2000) *The Dinosaur Hunters: A true Story of Scientific Rivalry and the Discovery of the Prehistoric World*, Fourth Estate, London, p. 374.

Calder, N. (1972) *Restless Earth*, Futura Publications, London, p. 128.

Caldwell, B.R. (2005) *Geology in the Bible*, Exposure Publishing, Burgess Hill, UK, p. 103.

Chamber, R. (1846) *Vestiges of The Natural History of Creation*, 5th edn, John Churchill, London, p. 421.

Chernicoff, S., Fox, H., and Tanner, L. (2002) *Earth: Geologic Principles and History*, Houghton Mifflin Co, Boston, MA, p. 570.

Chernicoff, S. and Whitney, D. (2002) *Geology: An Introduction to Physical Geology*, 3rd edn, Houghton Mifflin Co, Boston, MA, p. 648.

Clarkson, E.N.K. (1998) *Invertebrate Palaeontology and Evolution*, 4th edn, Blackwell Science, Oxford, p. 452.

Condie, K.C. (1997) *Plate Tectonics and Crustal Evolution*, 4th edn, Butterworth Heinmann, Oxford, p. 282.

Condie, K.C. and Sloan, R.E. (1998) *Origin and Evolution of Earth: Principles of Historical Geology*, Prentice-Hall Inc., New Jersey, p. 498.

*Cope, J.C.W., Ingham, J.K., and Rawson, P.F. (eds) (1992) *Atlas of Palaeogeography and Lithofacies*. The Geological Society Memoir No. 13. Geological Society of London, Bath, p. 153.

*Coyne, J.A (2009) *Why Evolution is True*, Oxford University Press, Oxford, p. 309.

*Cracknell, B.E. (2005) *Outrageous Waves*, Phillmore & Co. Ltd, Chichester, p. 302.

*Cutter, A. (2003) *The Seashell on the Mountaintop: A Story of Science, Sainthood, and the Humble Genius who Discovered a New History of the Earth*, Arrow Books, London, p. 228.

Dixon, B. (1989) *From Creation to Chaos, Classic Writings in Science*. Basil Blackwell, Ltd, Oxford, p. 280.

Dowd, M. (2007) *Thank God for Evolution: How the Marriage of Science and Religion will Transform Your Life and Our World*, Plume Printing, New York, p. 411.

Doyle. P. and Bennett, M.R. (1998) *Unlocking the Stratigraphic Record: Advances in Modern Stratigraphy*, John Wiley & Sons, Chichester, P. 532.

*Doyle, P., Bennett, M.R., and Baxter, A.N. (1994) *The Key to Earth History: An Introduction to Stratigraphy*, 1st edn, John Wiley & Sons Ltd, London, p. 293.

Doyle, P., Bennett, M.R., and Baxter, A.N. (2001) *The Key to Earth History: An Introduction to Stratigraphy*, 2nd edn, John Wiley & Sons, Ltd., London, p. 293.

Drake, E.T. and Jordan, W.M. (eds) (1985) *Geologists and Ideas: A History of North American Geology*, The Geological Society of America, Inc., Boulder, CO, p. 525.

Edwards, W.N. (1976) *The Early History of Palaeontology*, British Museum (Natural History), London, p. 58.

Eicher, D.L. (1976) *Geologic Time*, Prentice/Hall International Inc., London, p. 150.

*Fagan, B. (2000) *The Little Ice Age*, Basic Books, New York, p. 246.

*Fagan, B. (2004) *The Long Summer: How Climate Changed Civilization*, Granta Books, London, p. 284.

Freeman, M. (2004) *Victorians and the Prehistoric: Track to a Lost World*, Yale University Press, New Haven, p. 310.

*Futuyma, D.J. (1995) *Science on Trial: The Case for Evolution*, Sinauer Associates Inc, Massachusetts, p. 287.

*Garner, P. (2009) *The New Creation: Building Scientific Theories on a Biblical Foundation*, Evangelical Press, Darlington, UK, p. 300.

Garwood, C. (2007) *Flat Earth: The History of an Infamous Idea*, Macmillan, London, p. 436.

Gingerich, O. (2004) *The Book Nobody Read: Chasing the Revolutions of Nicolaus Copernicus*, Penguin Books, New York, p. 306.

Gish, D.T. (1974) *Evolution the Fossils Say No!* Creation-Life Publishers, San Diego, p. 129.

Godfrey, L.R. (1983) *Scientists Confront Creationism*. W.W. Norton & Company, New York, p. 324.

*Gould, S.J. (1987a) *Time's Arrow, Time's Cycle*, Penguin Books, London, p. 222.

Gould, S.J. (1987b), *An Urchin in the Storm*, Penguin Books, London, p 255.

*Gould, S.J. (1989) *Wonderful Life. The Burgess Shale and the Nature of History*, W.W. Norton & Company, New York, p. 347.

Gould, S.J. (1991) *Bully for Brontosaurus*, Penguin Books, London, p. 540.

Gould, S.J. (1993) *Eight Little Piggies*, Penguin Books, London, p. 479.

Gould, S.J. (1996) *Life's Grandeur: The Spread of Excellence from Plato to Darwin*, Jonathan Cape, London, p. 244.

Gould, S.J. (2001) *Rocks of Ages: Science and Religion in the Fullness of Life*, Jonathan Cape, London, p. 241.

*Hallam, A. (2004) *Catastrophes and Lesser Calamities: The Causes of Mass Extinctions*, Oxford University Press, Oxford, p. 226.

Helferich, G. (2004) *Humboldt's Cosmos: Alexander Von Humboldt and the Latin American Journey that Changed the Way We See the World*, Gotham Books, New York, p. 358.

Holmes, A. (1965) *Principles of Physical Geology*, 2nd edn, The Ronald Press Co, New York, p. 1288.

Howell, D.G. (1995) *Principles of Terrane Analysis: New Applications for Global Tectonics*, 2nd edn, Chapman & Hall, London, p. 245.

*Isaak, M. (2007) *The Counter-Creationism Handbook*, University of California Press, Berkley CA, p. 330.

*Jardine, L. (2003) *The Curious Life of Robert Hooke: The Man Who Measured London*, Harper Perennial, London, p. 422.

Johanson, D.C. and Edey, M.A. (1981), *Lucy: The Beginnings of Humankind*, Granada Publishing Limited, London, p. 409.

Kearey, P. and Vine, F.J. (1996) *Global Tectonics*, 2nd edn, Blackwell Scientific, Oxford, p. 333.

Leddra, M.J. (2008) *Turn and Burn: The Development of Coal Mining and the Railways in the North East of England*, University of Sunderland Press, Houghton-le-Spring. p. 96.

LeGrand, H.E. (1988) *Drifting Continents and Shifting Theories.* Cambridge University Press, Cambridge, p. 313.

*Lewis, C. (2002) *The Dating Game: One Man's Search for the Age of the Earth*, Cambridge University Press, Cambridge, p. 258.

*Lucas, E. (1989) *Genesis Today: Genesis and the Questions of Science.* Christian Impact, London, p. 160.

Lumsden, G.I. (ed.) (1994) *Geology and the Environment in Western Europe*, Oxford University Press, Oxford, p. 325.

Lyell, C. (1997) *Principles of Geology* (ed. J.A. Secord), Penguin Books, London, p. 472.

Manning, P.L. (2008) *Grave Secrets of Dinosaurs*, National Geographic, Washington DC, p. 316.

Mayor, A. (2000) *The First Fossil Hunters: Paleontology in Greek and Roman Times*, Princeton University Press, Princeton NJ, p. 361.

McCarthy, S. and Gilbert, M. (1994) *The Crystal Palace Dinosaurs: The Story of the World's First Prehistoric Sculptures*, The Crystal Palace Foundation, Crystal Palace, p. 99.

*McGowan, C. (2002) *The Dragon Seekers: The Discovery of Dinosaurs During the Prelude to Darwin*, Abacus, London, p. 270.

Monroe, J.S. and Wincander, R. (1994) *The Changing Earth: Exploring Geology and Evolution*, West Publishing Co, Minneapolis, p. 731.

Moore, R. (1961) *Man, Time and Fossils: The Story of Evolution.* Alfred A. Knopt, New York, p. 382.

Northdurft, W., Smith, J., Lamanna, *et al.* (2002) *The Lost Dinosaurs of Egypt: The Astonishing and Unlikely True Story of One of the Twentieth Century's Greatest Paleontological Discoveries*, Random House Trade paperbacks, New York, p. 239.

*Oreskes, N. (1999) *The Rejection of Continental Drift: Theory and Method in American Earth Science*, Oxford University Press, New York, p. 420.

*Osborne, R. (1998) *The Floating Egg: Episodes in the Making of Geology*, Jonathan Cape, London, p. 372.

Palmer, D. (2003) *Fossil Revolution: The Finds that Changed our View of the Past*, Harper Collins Publishers, Ltd, London, p. 144.

*Palmer, D. (2005) *Earth Time: Exploring the Deep Past from Victorian England to the Grand Canyon*, John Wiley & Sons, Ltd, Chichester, p. 436.

Petto, A.J. and Godfrey, L.R. (eds) (2007) *Scientists Confront Creationism: Intelligent Design and Beyond*, W.W. Norton & Company Inc, New York, p. 463.

Pierce, P. (2006) *Jurassic Mary: Mary Anning and the Primeval Monsters*, Sutton Publishing Limited, Stroud UK, p. 238.

Porter, R. (1980) *The Making of Geology: Earth Science in Britain 1660–1815*, Cambridge University Press, Cambridge, p. 288.

*Prothero, D.R. (2007) *Evolution: What the Fossils Say and Why It Matters*, Columbia University Press, New York, p. 381.

*Raup, D.M. (1991) *Extinction – Bad Genes or Bad Luck?* W.W. Norton & Company, New York, p. 120.

*Repcheck, J. (2003) *The Man Who Found Time: James Hutton and the Discovery of the Earth's Antiquity*, Pocket Books, London, p. 247.

Robson, D.A. (1986) *Pioneers of Geology*, The Natural History Society of Northumbria, Newcastle, p. 73.

*Rudwick, M.J.S. (2005) *Bursting the Limits of Time: The Reconstruction of Geohistory in the Age of Revolution*. The University of Chicago Press, Ltd, London, p. 708.

Ryan, W. and Pitman, W. (1998) *Noah's Flood: The New Scientific Discoveries About the Event That Changed History*, Simon & Schuster, New York, p. 319.

*Schneiderman, J.S. and Allmon, W.D. (eds) (2009) *For the Rock Record: Geologists on Intelligent Design*, University of California Press, Berkeley CA, p. 261.

*Scott, E.C. (2004) *Evolution vs. Creationism: An Introduction*, University of California Press, Ltd, London, p. 272.

*Shubin, N. (2009) *Your Inner Fish: The Amazing Discovery of our 375-million-year-old Ancestor*, Penguin Books, London, p. 237.

*Snoke, D. (2006) *A Biblical Case for an Old Earth*, Baker Publishing Group, Grand Rapids, MN, p. 223.

Standish, D. (2006) *Hollow Earth: The Long and Curious History of Imagining Strange Lands, Fantastical Creatures, Advanced Civilisations, and Marvellous Machines Below the Earth's Surface*, Da Capo Press, Cambridge MA, p. 303.

Sterelny, K. (2007) *Dawkins vs. Gould: Survival of the Fittest*, Icon Books, Cambridge, p. 205.

Summerfield, M.A. (1991) *Global Geomorphology*. Longman Scientific & Technical, Harlow UK, p. 537.

Tarbuck, E.J. and Lutgens, F.K. (2002) *Earth: An Introduction to Physical Geology*, 7th edn, Prentice Hall, New Jersey, p. 669.

Tassy, P. (1991) *The Message of Fossils*, McGraw-Hill Inc, New York, p. 163.

Thenius, E. (1973) *Fossils and the Life of the Past*, The English University Press, Ltd, London, p. 193.

Vail, T. (2003) *Grand Canyon: A Different View*. Master Books, Green Forest AZ, p. 103.

Van Angel, T.H. (1994) *New Views of an Old Planet: A History of Global Change*, Cambridge University Press, Cambridge, p. 430.

*Walker, G. (2003) *Snowball Earth*, Three Rivers Press, New York, p. 269.

Wells, J. (2006) *The Politically Incorrect Guide to Darwinism and Intelligent Design*, Regnery Publishing Inc, Washington DC, p. 273.

Whitcomb, J.C. and Morris, H.M. (1961) *The Genesis Flood*, Presbyterian & Reformed, Philadelphia NJ, p. 212.

Whorton, M.S. (2005) *Perils in Paradise: Theology, Science, and the Age of the Earth*, Authentic Media, Waynesboro GA, p. 233.

Wincander, R. and Monroe, J.S. (1993) *Historical Geology: Evolution of the Earth and Life Through Time*, 2nd edn, West Publishing Company, Minneapolis, p. 640.

*Winchester, S. (2002) *The Map that Changed the World: A Tale of Rocks, Ruin and Redemption*, Penguin Books, Ltd, London, p. 338.

Winchester, S. (2004) *Krakatoa: The Day the World Exploded 27 August 1883*, Penguin Books, London, p. 432.

Winchester, S. (2005) *A Crack in the Edge of the World*, Penguin Books, London, p. 412.

Woodcock, N. and Strachan, R. (eds) (2000) *Geological History of Britain and Ireland*, Blackwell Scientific, Oxford, p. 423.

Young, D.A. (1995) *The Biblical Flood*, William B. Eerdmans Publishing Co, Michigan, p. 327.

Index

Bold type denotes figures or tables.

Time Matters: Geology's Legacy to Scientific Thought, 1st edition. By Michael Leddra. Published 2010 by Blackwell Publishing Ltd.